Communicating
User Experience

STUDIES IN NEW MEDIA

SERIES EDITOR: JOHN ALLEN HENDRICKS,
Stephen F. Austin State University

This series aims to advance the theoretical and practical understanding of the emergence, adoption, and influence of new technologies. It provides a venue to explore how New Media technologies are changing the media landscape in the twenty-first century.

TITLES IN SERIES:

Communicating User Experience

Applying Local Strategies Research to Digital Media Design

Edited by Trudy Milburn

LEXINGTON BOOKS
Lanham • Boulder • New York • London

Published by Lexington Books
An imprint of The Rowman & Littlefield Publishing Group, Inc.
4501 Forbes Boulevard, Suite 200, Lanham, Maryland 20706
www.rowman.com

6 Tinworth Street, London SE11 5AL

British Library Cataloguing in Publication Information Available

Library of Congress Cataloging-in-Publication Data

Communicating user experience : applying local strategies research to digital media design / edited by Trudy Milburn.
pages cm. — (Studies in new media)
Includes bibliographical references and index.
ISBN 978-1-4985-0613-7 (cloth) — ISBN 978-1-4985-0615-1 (pbk.) —
    ISBN 978-1-4985-0614-4 (electronic) 1. Human-computer interaction. 2. Digital media. I. Milburn, Trudy.
QA76.9.H85C6548 2015
004.01'9—dc23
                                                        2015011634

To
June E. Christian

Contents

Transcription Notation

[ ]	*Brackets show overlapping talk*
=say	*Equal signs show latching (no interval) between utterances*
(.)	*A period inside parentheses shows a pause.*
	Longer pauses are indicated by adding periods
but-	*A dash shows sharp cutoff of speech*
better	*Underlining indicates emphasis*
NEVER	*Capital letters show talk that is noticeably louder than the surrounding talk*
°what is°	*Degree signs indicate talk that is noticeably more quiet than the surrounding talk*
>fast<	*"Less than" and "greater than" signs indicate talk that is*
<slow>	*noticeably faster or slower than the surrounding talk.*
ple:ase	*A colon indicates an extension of the sound or syllable that it follows.*
	Multiple colons indicate longer extension
↑↓	*Arrows pointing upwards and downwards indicated marked rising and falling shifts in intonation in the talk immediately following.*
( )	*Unclear speech or word; Words between parentheses represent the best guess of a stretch of talk which was difficult to hear*
((cough))	*Double parentheses with italicized content enclose transcriber's descriptions of sounds or other features of the talk/scene*
(1.5)	*Numbers between parentheses indicate length of pauses in seconds and tenths of seconds*

Adapted from the work of Gail Jefferson, from Atkinson, J. M., & Heritage, J. (1984). *Structures of social action: Studies in conversation analysis.* Cambridge: Cambridge University Press.

Preface

Trudy Milburn

This book is the direct result of ongoing scholarship and intentional scholarly discussion, each phase of which has purposefully sought to articulate contemporary contributions of the ethnography of communication tradition to local strategies research (LSR), applied communication research and design.

In the summer of 2012, a group of scholars working in the ethnography of communication met at Creighton University in Omaha, Nebraska. Organized by Donal Carbaugh and Jay Leighter, the group assembled to celebrate the anniversaries of influential work in the field including, especially, Dell Hymes's (1972) "Models of the Interaction of Language and Social Life" and Gerry Philipsen's (1992) *Speaking Culturally: Explorations in Social Communication*. The majority of scholars who attended were intellectual descendants of Gerry Philipsen and what has since been called "The Washington School" of the ethnography of communication (Leeds-Hurwitz 2010; Carbaugh 2010, 1995). Conference attendees traveled from thirty-one universities/organizations and eight countries to participate in the three-day conference with the Creighton community, celebrate past achievements, and map "ways forward" in the intellectual tradition. Many research relationships were re-kindled including, especially, scholars trained by graduates of the Washington School in now established ethnography of communication programs at the University of Haifa (Israel), University of Massachusetts–Amherst, and University of Iowa. Discussions centered on shared methodological practices and new applications for ethnography of communication research. Among the twenty-eight papers presented, research featured in the book was first shared publicly, including Molina-Markham et al.'s chapter in an address by Donal Carbaugh on in-car communication.

Stemming from a conversation at that conference, Leah Sprain and David Boromisza-Habashi from the University of Colorado Boulder edited a special issue of Journal of Applied Communication Research titled *Ethnographers of Communication in Applied Communication Research*. Conversations at the Creighton conference helped set the trajectory for the special issue which ultimately provided a forum for scholars to collaborate on papers about the application of ethnography of communication. The special issue included a piece significant to this volume by Jay Leighter, Lisa Rudnick, and Theresa J. Edmonds about principles relating to design.

Building on the momentum of the conference, several members of this same group and a few new faces came together the fall of 2013 for an NCA pre-conference organized by Michelle Scollo, from the College of Mount Saint Vincent, titled, "Talking Technology: New Connections in the Ethnography of Communication and Technology." This pre-conference was focused specifically on the theme of ethnography of communication and technology in honor of *Talking American*, as the first work focused entirely on electronic media. Several authors contained in this volume presented research during that pre-conference including Tabitha Hart, James Leighter, Lauren Mackenzie, Trudy Milburn, and Todd Sandel.

Most recently, Leighter organized a gathering at a Blue Sky workshop at ICA in May of 2014 for like-minded scholars to discuss the concept of LSR. This concept, originated by Gerry Philipsen and Lisa Coutu at the University of Washington, has since appeared in research articles including the JACR special issue. Upon Philipsen's retirement, the question became how and where to continue to host this type of research. The outcome from this conversation has been reframed in the introduction to this volume in an effort to begin to ground the notion in published scholarship and provide the parameters for this type of research.

Even though this volume did not include all of the participants at these gatherings, we are extremely grateful for the conversations and feedback that everyone provided.

Introduction

Local Strategies Research: Application to Design
James L. Leighter and Trudy Milburn

This book provides an opportunity to reflect upon design strategies based on actual user-interactions analyzed through established Language and Social Interaction (LSI) methodologies. In so doing, we hope to contribute to a call for the "communication design enterprise" based on the fundamental assumptions that communication design is a "natural fact about communication" (Aakhus and Jackson 2005, 413) and designs reveal assumptions about how communication can and should work (Aakhus and Jackson 2005; Aakhus 2007). Each chapter will explore different perspectives based on the confluence of distinctive but overlapping fields of research, activity, and inquiry: communication scholars involved in local strategies research (LSR), human-computer interaction (HCI) research into social interaction (Carroll 2013; Rogers, Sharp, and Preece 2011), and both product and service design (Kimbell 2011, 2012; Buchanan 1992; Margolin and Buchanan 1995; Norman 2002).

This introductory chapter addresses our research assumptions through a brief discussion of our shared research perspective, an inquiry into terminology made necessary by the trans-disciplinary nature of the research, and the establishment of critical definitional points of departure.

LOCAL STRATEGIES RESEARCH

The assumption that communication research may have positive influence on the design process is grounded, in part, in the empirical research presented in the book. The assumption is also an extension of a newly forming research perspective: *Local Strategies Research* (LSR). The term

LSR was first introduced by Philipsen in conference presentations, personal communication, and an unpublished paper (2012) he posted as a web resource for our research collaborative entitled "Local Strategies Research: From Knowledge to Practice." In the paper, Philipsen suggests LSR is "an innovative approach to gathering data on local perceptions of needs, opportunities, and efficacious means for achieving constructive change." In these early articulations, Philipsen described LSR in terms that illumined a perspective for research more than a particular research methodology. Based on years of comparative ethnography of communication research, some of which was explicitly conducted as applied ethnography of communication research, Philipsen articulated in his paper what he called the "core wisdom" of LSR:

> That paying attention to what people do, particularly in their moments of acting so as to accomplish something socially, can be crucial to finding out something about what they take to be the problematic or the possible in social life and to finding out their notions of how to contend with the problematic and realize the possible in the circumstances in which they find themselves.

Inspired by Philipsen's initial formulation of LSR, many scholars have begun to use the term suggesting a variety of ways in which their research is guided by LSR principles. In conference presentations, competitive grant applications, conceptual essays, and calls for social research, the term LSR has been used to bolster claims about research assumptions, perspectives, methods, and theories, particularly in instances when the goal was to intervene in some way to improve the daily life of a particular community. Philipsen characterizes LSR in intervention circumstances as questions researchers should consider when making the distinction between the activities of demonstration and conversation. The former presumes that those intending to intervene in a community have developed specialized knowledge that is available for relatively simple transmission to a community. The goal is to provide the most compelling demonstration of this knowledge while articulating possible positive outcomes for the community in question. While demonstration may include cursory and short-lived efforts to understand, observe, and/or talk with community members, such efforts are meant to bolster conditions for the demonstration.

A predisposition toward conversation, in contrast, calls for a delay in immediate intervention, allowing time for more concentrated and, perhaps, long-lasting looking and listening in the community. The purpose is to understand the terms, meanings, premises, and practices that constitute daily life in the community and, then, to ask of this knowledge what value it has for shaping, guiding, or even preventing an intervention.

LSR, as it was first articulated by Philipsen, was deeply indebted to the ethnography of communication tradition. Ethnography of communication research has been a method employed across disciplines including, prominently, anthropology and linguistics. Over the years, many scholars have used the basic methodological building blocks proposed by Dell Hymes (1962, 1972) and elaborated by Philipsen to describe the social codes (1992, 1997) members rely upon when they interact. Another variation of ethnography of communication has been developed through cultural discourse analytic frameworks (Carbaugh 2007; Carbaugh, Gibson, and Milburn 1997; Scollo 2011) that can be used to make sense of a diverse set of practices and user experiences. Each chapter in this text relies upon one application of cultural discourse theory to a technological context or problem that necessitated obtaining additional information about actual users interacting with or through digital media.

LSR's beginnings were also deeply influenced by design thinking. Common ground exists, then, between the LSR researcher predisposed to look and listen before intervening, and the designer who is interested in developing knowledge of users as part of the design process. For instance, Philipsen directly references design research seeking solutions to peace and security in Ghana as part of the Security Needs Assessment Protocol (SNAP) project of the United Nations Institute for Disarmament Research (UNIDIR; see particularly Miller and Rudnick 2008) as influential in his thinking about the nature of LSR. Philipsen also drew from his experience at the Glen Cove Conference on Strategic Design and Public Policy, co-sponsored by the University of Washington Center for Local Strategies Research, UNIDIR and the Said Business School at the University of Oxford. A summary of key findings from that conference report relevant for this book include the following:

- Finding the essence of a problem takes time, and it needs attention and resources so that the design process can set about the right tasks in the right way.
- Conversations between cultural research specialists (with expertise in ethnography and interpretive methods) and designers seem both novel and hold high promise.
- Design can direct cultural research towards application solutions, whereas cultural research can make design more reflective about its assumptive base and premises for practice.
- Cultural research can produce knowledge that can be assembled as a resource in design activities.
- Design research can produce insights and artifacts that inspire and mobilize participants in design processes.

Whereas these findings were developed in direct response to problems of policy, our book shows implications for them in design more broadly and, in the chapters that follow, in the specific domain of digital media. Thus, this book is an opportunity to interrogate the concept of LSR and its relationship to design in a more nuanced way. It is an opportunity to take up the concept of LSR as a guiding set of assumptions and an attitude toward social research in order to begin to refine and establish it as a field of scholarship.

DESIGN

Following Aakhus (2007), we begin from the premise that designers revere the values of creation, innovation, invention, intervention, redesign, and reflection. Two features of extant research in design suggest a prominent role to be played by communication scholarship. First, empirical examinations of the design process from a communication perspective often conclude with a call for a reconfiguration of current communication practice including reconfiguring the design process, or recalibrating the way in which interaction analysis between products/services and people occurs. Second, there is tremendous potential for co-designing material artifacts and services that are much more responsive to local, cultural contexts because of the way communication research can infuse empirical, data-driven knowledge of communities into the design process. By this we mean, the construction of products and services can be improved with better methodologies for the discovery of how humans will interact with one another as they experience or interact with the material artifact itself. Put simply, we propose that there is great potential in the examination of cultural notions when improving design process and designs.

Design processes vary as widely as the products and services designers prototype, develop, and deploy. Most design processes, however, can be described in terms of phases of problem-identification, investigation of users through research, prototyping, development and deployment. Our main concern is to neither reify design processes by grossly oversimplifying them nor to take them on face-value without critique. Nonetheless, we recognize that there are established principles for the design process as is evidenced in, for example, the educational practices in design schools. Instead, our main concern is to raise the questions of when and how communication research can be influential in the design process.

The chapters in this book show promise for positively influencing the design process in three key ways: 1) in the discovery and articulation of local problematics, including the articulation of the relationships between designs and people (or personas); 2) in the prototyping of new designs,

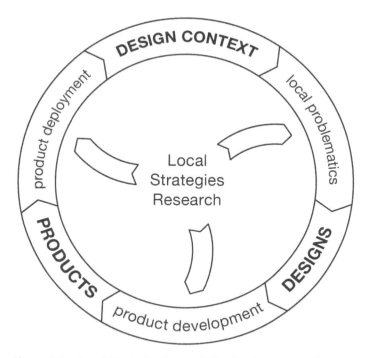

Figure 0.1. Local Strategies Research Applications for Design Process. Theresa Edmonds, illustrator.

especially with regard to typical evaluation processes of user-interaction with designs; and 3) in the examination of the cultural or communicative influence of designs as they are deployed in particular communities of use.

Because of the nature of our cases, we choose to emphasize in this book these capacities for LSR in product design, but believe they would apply to service design or other types of design intervention as well.

For discovering local problematics, LSR places emphasis on early and concentrated conversations with those for whom the designs are intended. By focusing on the everyday life-world of these people answers to several questions may come into focus including: how are "users" conceptualized not just with respect to the potential design but richer terms of communication practices, identity, and social relationships? And, if identifying the problem or circumstances the designs are intended to address were left to the people who will eventually use the design, how would they themselves describe the problem?

For influencing product development, local strategies research holds promise for shaping developmental iterations of prototypes through collections and analyses of locally situated data. Designers employ a vast array of tools and processes for prototype development. The question becomes how might these tools and processes be improved through the discovery of local practice during these phases of development?

Once a product has been deployed, what is there to learn about its use? Local strategies research is useful here not just in the reshaping and reconfiguring of design improvements, but in rethinking the parameters of the design context for new products and applications. Building from these particular moments in which LSR can contribute to design process, there are at least three main contributions LSR can make to design methods, including

1. orientation to communication as the theoretical equivalent to interaction;
2. provision of a means to see and hear micro (conversational) and macro (relational) interactional processes that are either rule-based or normatively sequenced;
3. cross-cultural comparisons that provide information about the breadth and depth of differences between communicative practices.

In essence, we suggest LSR has the potential to help designers develop cultural competence in the communities for whom they are designing. As Carbaugh and Lie (2014) point out, cultural competence is not an abstraction of language or universal understanding located somewhere outside of everyday situations and activity. Rather, "what is competent must be understood to vary by social situation and means of communication, by cultural meaning and interactional scene" (74). We suggest that the inclinations of designers to learn from users as part of the design process can be enhanced by a deeper understanding of what makes communication, for a particular community in particular scenes, appropriate and effective. What is required to do so is a commitment to exploring the life-worlds of users in a systematic way.

"USERS" AS CULTURAL PERSONAS

In this book, one particular component of design discourse merits special attention. When designing products and services, as digital media or otherwise, designers' preferred term for categorizing the people who will experience their design is user. Part III of the book addresses this issue specifically, but it is useful here to introduce our perspective to set the stage for conceptual clarity on this important term. As with design processes, our goals are neither to reify use of this term, nor to take it up in our studies unreflectively. We acknowledge widespread use of this term in the field of design, especially in the sub-fields that focus on "user-oriented design," and we are mindful of the ways in which this term has become contested in design. Regarding the latter point, we follow Murray's (2012) treatise on problems with design terminology when she articulates concerns about "user" having a limited terminological capacity. She prefers the term "interactors" who can be "understood in larger contexts than as the tool users and task performers. They also make complex judgments about what they want to know, what they want to do, and where they want to go in the digital realm" (62).

This book contributes to a conversation on concepts and terminology for persons by arguing that there is an extraordinary opportunity for considering the ways local strategies research can contribute to a refashioned notion of users, particularly in the field of human-computer interaction (HCI). Typically in design, notions of users are developed as an inspiration for the design process but often fade into the background during the creative design process, resulting in a gap between intentional inquiry into the lives of users and the final designs (Leighter, Rudnick, and Edmonds 2013). In contrast, our LSR approach seeks to develop meaningful, empirical, and robust notions of personas, perhaps wildly expanding notions of who "users" are and what they can do. As is mentioned above, we see opportunities for expanding notions of users in the investigation of local problematics, product development, and product evaluation for improvement. Such notions have the potential to persist and guide designers through the full design process and, thus, provide a permanent empirically driven link between investigations of people and the realized products and services they will experience.

In order to preview this potential, we draw from cultural discourse theory (CDT), a theory that is not ubiquitous but pervades much of the book. As explained in Carbaugh, Gibson, and Milburn (1997),

> The meanings of cultural discourses "of the symbols, acts, forms, rules" consist in basic premises about being (identity), doing (action), relating (social relations), feeling (emoting), and dwelling (living in place). (22)

What CDT illuminates is the potential to conceptualize, investigate, describe, and analyze personhood (with respect to designs) in complex ways, ways that mirror the everyday activity of humans in action. This is not to suggest that all dimensions of personhood are equally relevant or salient in everyday activities, but rather that a focus on these dimensions opens up the possibility for discovery of personhood as multifaceted.

Just as there is conceptual overlap between LSR and the design process, so too is there conceptual overlap between a notion of persona developed in CDT and design. For example, Alan Kay (1990), in part responsible for the design of Apple's Macintosh computer language and object-oriented programming, describes communication theories and principles as among the origins of his thinking, including Marshall McLuhan's premise about the "medium" and symbols from semiotics. Like most HCI practitioners, Kay turned his attention from these communication theories to focus on the psychological aspects of a "thinking" individual. This focus on user personalities is manifest most clearly in the creation of personas (Maguire 2001; Cooper 2004; Dantin 2005; Caddick and Cable 2011). Personas are constructed as an amalgam represented by a typified individual, who has a particular role and job tasks, and who may be characterized with personality traits and emotions. While designers have found the creation of personas a useful way to reconsider a product from a particular set of needs based on an individual (Ayoma 2007), their characterization often divorces a user's persona from her goals and the scenario in which she finds herself. If we turn to communication scholars, such as those writing about personhood in *Communication Yearbook* (Deetz 1994), as well as some discursive psychologists (see Potter and Wetherell 1987; Edwards and Potter 1992), we find a move away from a view of persons as unitary, ontological beings to a broader focus on the way personhood and social action are accomplished in and through social interaction itself. We raise the issue of personas because beliefs about personhood greatly inform design. Our approach is intended to be expansive and complementary to established design processes.

DIGITAL MEDIA

Before introducing the cases, one final point needs to be made about the technologies examined in this book. Taking a step back into the history of information theory, Shannon and Weaver created a model for communication that was deeply influenced by the dominant communication technology of the day: the telephone. In their model, a sender encoded a message to send through a medium (such as a telephone) so that a receiver could decode the message. Their goal was to reduce the impact of noise to ensure the clarity of the messages sent between senders and receivers. Since the time of Shannon and Weaver's theory, mass communication research continued to bleed into interpersonal research. What was once considered a realm of study based on a medium made for broadcasting messages to many (that is, television) morphed into a single device that can be used to interact with one or many (that is, a smartphone). Traditional communication research was based on number-of-speakers and number-of-hearers distinctions for the first several decades such research was conducted. Over time, these distinctions have become less useful for describing communicative technologies, on the one hand, and engaging in research of communicative technologies, on the other.

This book includes research that is based upon a range of technologies within different scenarios including three that refer to devices that transcend place (including mobile phones, an in-car interface, and a social media site called WeChat) and three data-based chapters focused on online learning (chapters 2, 4, and 7). One chapter refers to interaction in a meeting conducted virtually (chapter 4). Rather than speaking of these as technologies, we will follow Murray's (2012) lead and refer broadly to the impact (and use) of communication theories and methods on digital media. Murray believed the term digital media was more descriptive than "new media" because it includes both an array of "computer-based artifacts" and applications "aimed at complex cultural communication" as well as whatever is invented in the future. She thought that "the vagueness of the term encourage[d] sloppy thinking about design by suggesting that novelty [was] the salient property of these phenomena" (8). She further argues that "calling objects made with computing technology 'new' media obscures the fact that it is the *computer that is the defining difference not the novelty*" (8). Therefore she advocates including digital artifacts as part of "a single new medium that is created by exploiting the representational power of the computer" (8). By so doing, Murray says we can focus properly on four dimensions: "procedural, participatory, spatial, and encyclopedic affordances" (9).

In order to establish a more productive starting point for our research, we choose to follow Murray's lead and reference the term *digital media* to characterize the studies in our book. Doing so sufficiently and pragmatically captures the quintessential essence of the communicative situations that are the foci of our studies. Use of this term provides theoretical congruence with assumptions made in the ethnography of communication tradition. Hymes (1972) deftly articulates structural features of communication that are useful theoretical and methodological starting points for the discovery and delimitation of particular kinds of communicative activity. For him, a speech (communicative) situation is one that is marked by the presence or absence of speech (communication) and, more importantly, is "in some recognizable way bounded or integral" (56) to those who participate in the activity. Because our book examines a vast array of communicative activities, we choose to highlight what makes the cases hang together. It is the communicative interaction with, in, and through digital media in a wide variety of forms and manifestations. The degree to which the participants in our cases are beholden to the design of a particular technology varies widely. But in each case, the digital nature of the technology present in the interactions plays a crucial role in understanding the communicative activity featured in the analysis.

Taking these terms and concepts into consideration, each section of the book addresses one central concern, but one can easily recognize overlap between and among the chapters. By applying LSR and building from CDT, the chapters show a commitment to answering the following questions of each case:

- What are users doing (which activity or tasks are they performing) and towards what ends?
- How do users relate to one another (and the digital medium)?
- Who do users presume to be when they interact? Do different person positions or cultural premises come into play?

In sum, we begin this volume in the way Carbaugh (1994) advocates:

> One begins, then, not by assuming a typology of persons, relations, or actions as something prior to discursive action but by assuming that activities of positioning indeed take place in discourse and then investigating the nature of that activity in that discourse through a conceptual framework. (182)

REFERENCES

Aakhus, Mark. "Communication as Design." *Communication Monographs* 74, no. 1 (2007): 112–117. doi:10.1080/03637750701196383.

Aakhus, Mark, and Sally Jackson. "Technology, Design, and Interaction." In *Handbook of Language and Social Interaction,* edited by Kristine L. Fitch and Robert E. Sanders, 411–36. Mahwah, NJ: Erlbaum, 2005.

Aoyama, Mikio. "Persona-Scenario-Goal Methodology for User-Centered Requirements Engineering." *15th IEEE International Requirements Engineering Conference* (2007): 185–194. DOI 10.1109/RE.2007.50.

Buchanan, Richard. "Wicked Problems in Design Thinking." *Design Issues* 8, no. 2 (1992): 5–21.

Caddick, Richard, and Steve Cable. *Communicating the User Experience: A Practical Guide for Creating Useful UX Documentation.* West Sussez, UK: Wiley & Sons, 2011.

Carbaugh, Donal. "Personhood, Positioning and Cultural Pragmatics: American Dignity in Cross-Cultural Perspective." In *Communication Yearbook,* volume 17, edited by Stanley A. Deetz, 159–186. New York: Routledge, 1994.

Carbaugh, Donal. "Cultural Discourse Analysis: Communication Practices and Intercultural Encounters." *Journal of Intercultural Communication Research* 36, no. 3 (2007): 167–182. doi:10.1080/17475750701737090.

Carbaugh, Donal, Timothy Gibson, and Trudy Milburn. "A View of Communication and Culture: Scenes in an Ethnic Cultural Center and a Private College." In *Emerging Theories of Human Communication*, edited by Branislav Kovacic, 1–24. Albany, NY: SUNY Press, 1997.

Carbaugh, Donal, and Sunny Lie. "Competence in Interaction: Cultural Discourse Analysis." In *Intercultural Communication Competence: Conceptualization and its Development in Cultural Contexts and Interactions*, edited by Xiaodong Dai and Guo-Ming Chen, 69–81. Cambridge: Cambridge Scholars Publishing, 2014.

Carroll, John M. "Human Computer Interaction—Brief Introduction." In *The Encyclopedia of Human-Computer Interaction*, edited by Mads Soegaard and Rikke Friis Dam. 2nd edition. Aarhus, Denmark: The Interaction Design Foundation, 2013. Accessed February 6, 2015, http://www.interactiondesign.org/encyclopedia/human_computer_interaction_hci.html.

Cooper, Alan. *The Inmates Are Running the Asylum: Why High Tech Products Drive Us Crazy and How to Restore the Sanity*. Indianapolis, IN: Sams Publishing, 2004.

Dantin, Ursula. "Application of Personas in User Interface Design for Educational Software." In *Australasian Computing Education Conference. Conferences in Research and Practice in Information Technology*, edited by A. Young and D. Tolhurst, 42, 239–247. Newcastle, Australia: 2005.

Deetz, Stanley, ed. *Communication Yearbook*, volume 17. New York: Routledge, 1994.

Edwards, Derek, and Jonathan Potter. *Discursive Psychology*. Thousand Oaks, CA: Sage, 1992.

Hymes, Dell. "The Ethnography of Speaking." In *Anthropology and Human Behavior*, edited by Thomas Gladwin and William Sturtevant, 13–53. Washington, DC: The Anthropological Society of Washington, 1962.

Hymes, Dell. "Models for the Interaction of Language and Social Life." In *Directions in Sociolinguistics: The Ethnography of Communication*, edited by John Gumperz and Dell Hymes, 35–71. New York: Basil Blackwell Inc., 1972.

Kay, Alan. "User Interface: A Personal View." In *The Art of Human-Computer Interface Design*, edited by Brenda Laurel and S. Joy Mountford, 191–207. Reading, MA: Addison-Wesley Professional, 1990.

Kimbell, Lucy. "Rethinking Design Thinking: Part 1." *Design and Culture* 3 (2011): 285–306.

Kimbell, Lucy. "Rethinking Design Thinking: Part 2." *Design and Culture* 4 (2012): 129–148.

Leighter, James L., Lisa Rudnick, and Theresa J. Edmonds. "How the Ethnography of Communication Provides Resources for Design." *Journal of Applied Communication Research* 41, no. 2 (2013): 209–15.

Maguire, Martin. "Methods to Support Human-Centred Design." *International Journal of Human-Computer Studies* 55 (2001): 587–634. doi:10.1006/ijhc.2001.050.

Margolin, Victor, and Richard Buchanan, eds. *The Idea of Design: A Design Issues Reader*. Cambridge, MA: MIT Press, 1995.

Miller, Derick, and Lisa Rudnick. *The Security Needs Assessment Protocol: Improving Operational Effectiveness through Community Security*. Report. New York and Geneva: United Nations Publications, 2008.

Murray, Janet H. *Inventing the Medium: Principles of Interaction Design as a Cultural Practice*. Cambridge, MA: The MIT Press, 2012.

Norman, Don. *The Design of Everyday Things*. New York: Basic Books, 2002.

Philipsen, Gerry. *Speaking Culturally: Explorations in Social Communication*. Albany, NY: State University of New York Press, 1992.

Philipsen, Gerry. "A Theory of Speech Codes." In *Developing Communication Theories*, edited by Gerry Philipsen and Terrance L. Albrecht, 119–56. New York: State University of New York Press, 1997.

Philipsen, Gerry. "Local Strategies Research: From Knowledge to Practice." Unpublished paper, 2012.

Potter, Jonathan, and Margaret Wetherell. *Discourse and Social Psychology: Beyond Attitudes and Behaviour*. Thousand Oaks, CA: Sage, 1987.

Rogers, Yvonne, Helen Sharp, and Jenny Preece. *Interaction Design: Beyond Human-Computer Interaction*. West Sussex, UK: John Wiley & Sons, 2011.

Scollo, Michelle. "Cultural Approaches to Discourse Analysis: A Theoretical and Methodological Conversation with Special Focus on Donal Carbaugh's Cultural Discourse Theory." *Journal of Multicultural Discourses* 6, no. 1 (2011): 1–32.

I

COMMUNICATIVE ACTS AND PRACTICES
Trudy Milburn

Developers of digital media create interfaces that are activity-based. Their goal is for users to be able to perform a specific task with the new device or technology. In order to help developers and designers learn more about the tasks people accomplish when using new tools, researchers in this section address the question, "which activities are being accomplished when people interact with a specific digital medium in a particular situation?" The chapter authors provide detailed analyses to address this question with the hope that if we can recognize what tasks people are actually performing when they interact with digital media, that should help us to better create and modify existing systems and devices.

In order to address the question of which activities are being accomplished, the authors draw upon Dell Hymes's SPEAKING framework developed in the 1960s. Three of Hymes's SPEAKING categories: acts, ends, and norms (specifically norms for action and actions directed towards specific ends) are the cornerstones of the work recounted within this section. Hymes's acts category has much in common with Austin and Searle's work differentiating speech acts. Austin and Searle's early formulations of speech acts were based upon the premise that individual speakers are engaged not only in particular types of speaking activities, but that through conversation, speakers accomplish particular ends. Searle (1969) built upon Austin's (1962) three categories to postulate a set of five categories of speech acts including: assertives, commissives, expressives, directives, and declaratives. While these categories are useful for recognizing the ways that speech is used to perform and accomplish particular ends, these categories rely upon an analyst to separate both a

1

speaker's act into a single category as well as the intention of the speaker about the goal he or she seeks to accomplish.

This first section builds upon and transcends the original formulation of speech acts. Rather than beginning with a finite set of speech act categories, an LSR analysis (drawing from Hymes's framework) proceeds by noticing specific actions that speakers perform. These actions may be similar to Searle's categories such as directives, but are more finely nuanced. When a speaker requests that a song be played on a radio, or a trainer directs a student to provide evidence of learning, these actions are recognizable in the moment of interaction and labeled with the terms that people used. In addition, it may be the case that any given action itself may actually be indicative of several kinds of activities being performed simultaneously. Thus, our analysis shows how acts are multi-functional.

LSR scholars recognize that it is useful to use the notion of speech acts as a way to designate the myriad types of activities that are occurring within a given scene, rather than trying to constrain the evidence they find into discreet categories without overlapping boundaries among different acts performed by speakers. The important point is that identifying communicative acts is a way of understanding how people shape events through speaking.

In addition to focusing on the actions that speakers perform, the two chapters in this section also (directly or implicitly) make claims about categories that are initially based on and interpreted theoretically as one step removed from the data. Within cultural discourse theory (CDT), interpretive moves are made to postulate "practices" in ways that are distinct from single communicative acts. Using CDT, researchers examine data to formulate patterns of actions that may be enacted by members of a cultural group, actions that necessarily transcend individual speakers. With this summative analytic move, researchers in this tradition are concerned more with faithfully representing the speech acts that participants both refer to and enact within naturally occurring data.

Communicative practices have been defined as "a pattern of situated, message endowed action that is used in a scene(s)" (Carbaugh, Gibson, and Milburn 1997, 6). One may be tempted to conclude that every communicative act is "message endowed." Scollo (personal communication 2014) explains that some scholars prefer to examine the broader categories of communicative actions, because they convey meaning, rather than a more constricted view of practices that specifically describe patterned actions. Considered another way, a particular member of a culture may engage in an activity or perform an action one time, but it is not necessarily the case that the same action has been performed by other members of the culture—and thus constitutes a pattern. One powerful way to recognize culture at work is through noticing patterned practices, rather than simply focusing on discrete actions. Scollo summarizes that cultural discourses are comprised of practices, rules/norms, and their meanings (2011, 14). By focusing on patterned practices researchers can more easily examine normative ways of interacting.

For academic researchers and practitioners, there is great utility in beginning with communicative activities that may or may not lead to patterned practices. First, it is a way to learn more about the behavioral norms that shape users' expectations of any particular interaction. User expectations include two perspectives, both for how the interaction will be structured and progress as well as how normative breaches will be handled. In fact, scholars have long studied proper ways to account for one's untoward actions as well as ways to redress grievances (Scott and Lyman 1968). Often the ways interlocutors choose to move forward within an interaction are based on cultural patterns and rituals that should be carefully followed.

While designers provide functionality for a primary task as well as peripheral tasks, people may be enacting a number of activities that either relate directly to those tasks or are tangential to them. For instance, imagine someone driving to a destination while listening to music. The act of driving and listening are both being engaged in. By posing one of these as a primary goal, such as arriving at a destination, and the other as a secondary goal, such as finish listening to a particular song before disembarking from a car, developers may miss the way these acts are intertwined, and more or less salient moment-by-moment. In fact, in this example, drivers may also achieve other goals as a result of their actions such as relaxing or reminiscing by chatting with other passengers within the car. In the following chapter, Molina-Markham, van Over, Lie, and Carbaugh delve into these questions by narrowing their focus on how drivers establish the beginning and ending to requests within in-car talk. This important work draws upon what we know about interpersonal conversations and compares that to communication with a new digital medium.

Rather than asking what users think they are up to when they engage with digital media, both chapters that follow explore interactional sequences that results from normative breaches, different kinds of interactional failure. As Goffman and Garfinkel have explained, norms are often more easy to observe when they are broken. A common way to discover norms is to ask of a set of data, "given this practice, and the way people are doing it . . . what is it presumably that should/not be done?" (Carbaugh 2007, 178). While conversational participants notice and orient to breaches, positing a norm as a patterned practice is based on interpretive work done by the analyst.

In the two chapters in this section, Markham, van Over, Lie, and Carbaugh, and Hart focus specifically on user-interactions that illustrate what should/not be done. In the first, we learn more about reactions to an in-car prototype through the responses drivers make aloud while they are interacting with the new system. Whereas in the second, Hart explores the interaction between users while they interact with a new system for learning a second language. Within both scenarios, we find that the digital medium both enables and constrains the interaction, in both expected and unexpected ways. Hart explicitly describes the norms, premises, and rules for communication in her chapter (chapter 2) as "procedural knowledge." Her exploration of this term directly ties in concerns that designers have when creating digital media that can be easily learned.

The focus on actions and practices are all conducted within the realm of "interaction." Designers often refer to this as, "the combination of the procedural and participatory properties of the digital medium: the structures by which we script computers with behaviors that accommodate and respond to the actions of human beings" (Murray 2012, 12). The two chapters in this section illustrate the designer view—as that which connotes interaction between a single individual and her device or digital medium (primarily chapter 1). This is followed by a chapter that explores interaction between two primary participants, trainer and student, through the medium of the language learning website (chapter 2). Through her analysis, Hart help us to consider the ways that we juggle interaction with and through digital media when dyads are involved.

By investigating various acts performed when expectations are not met, we learn more about the ways people adapt. Humans have the unique ability for infinite flexibility with language during conversations, whereas digital media must be pre-programmed to respond or to act in specific ways. These next two studies raise questions about preferences for scripted interaction, calling into question the ways novel responses may be interpreted as normative breaches. While much human interaction is highly routinized, the expectations for how a conversation will unfold is also dependent upon nuanced features of the situation. Let us turn to some examples of that now.

REFERENCES

Austin, John L. *How to Do Things with Words*. Cambridge, MA: Harvard University Press, 1962.

Carbaugh, Donal. "Cultural Discourse Analysis: Communication Practices and Intercultural Encounters." *Journal of Intercultural Communication Research* 36, no. 3 (2007): 167–182. doi:10.1080/17475750701737090.

Carbaugh, Donal, Timothy Gibson, and Trudy Milburn. "A View of Communication and Culture: Scenes in an Ethnic Cultural Center and a Private College." In

Emerging Theories of Human Communication, edited by Branislav Kovacic, 1–24. Albany, NY: SUNY Press, 1997.

Murray, Janet H. *Inventing the Medium: Principles of Interaction Design as a Cultural Practice*. Cambridge, MA: The MIT Press, 2012.

Scollo, Michelle. "Cultural Approaches to Discourse Analysis: A Theoretical and Methodological Conversation with Special Focus on Donal Carbaugh's Cultural Discourse Theory." *Journal of Multicultural Discourses* 6, no. 1 (2011): 1–32.

Scollo, Michelle. Email message to the author, December 11, 2014.

Scott, Marvin B., and Stanford M. Lyman. "Accounts." *American Sociological Review* 33, no. 1 (1968): 46–62. Accessed February 6, 2015. http://links.jstor.org/sici?sici=0003-1224%281968O2%2933%3A1%3C46%3AA%3E2.0.CO%3B2-M.

Searle, John. *Speech Acts: An Essay in the Philosophy of Language*. Cambridge, UK: Cambridge University Press, 1969.

ONE

"OK, Talk to You Later"

Practices of Ending and Switching Tasks in Interactions with an In-Car Voice Enabled Interface

Elizabeth Molina-Markham, Brion van Over,
Sunny Lie, and Donal Carbaugh

While speaking with an in-car speech activated system during a research study, four participants laughed in response to the system's ending an interaction with the following phrase, "OK, talk to you later." One example is included below:

Participant 5
1:07:35

1. Participant: ((parks car and scrolls through contacts with touch)) OK. I can
2. scroll to touch, right. Anyway. Home.
3. System: OK, talk to you later.
4. Participant: ((laughs)) Hey that's kind of cute.

Participants' laughter in these instances could indicate that the in-car speech system's response was in some ways unanticipated (Coupland and Coupland 2001). Participants likely did not expect the system to reply at this point in the conversation with an acknowledgement that it would interact with them at a future time. Their expectations for how the encounter would end may not have been met. Researchers argue that people's expectations for the unfolding of an interaction are culturally shaped (Hymes 1962, 1972; Philipsen 1992; Carbaugh 2005, 2007). Norms of interaction guide what participants deem appropriate at each point in an encounter from the beginning to the end (Scollon and Scollon 1981). While in this particular case the use of the phrase, "OK, talk to you later," appears to have been at least to some extent appreciated by participants—as when Participant 5 notes "that's kind of cute"—there is also the possibility, which we will elaborate further below, of an interactant ending an encounter in an unanticipated way that is considered abrupt, rude, or inappropriate by a conversational partner. This failed ending may negatively influence a person's view of the other person (or the in-car system) with whom they are speaking, and it may even taint possible future encounters.

In designing a voice activated system, system designers draw on assumptions about the nature of the conversants with whom their system will interact and those users' expectations regarding dialogue flow. Given that premises regarding the nature of acting in this type of situation may vary between cultural communities, it is important for designers to take into account how their own premises could influence design in a way that may not be pleasing to users. As Murray (2012) writes "the design of digital objects is *a cultural practice like writing a book or making a film*" (1). She further explains, "In order to make truly intuitive interfaces, designers must be hyperaware of the conventions by which we make sense of the world—conventions that govern our navigation of space, our use of tools, and our engagement with media" (10). Therefore, it is essential to consider cultural premises when designing a digital system.

In this chapter, we analyze the ending of tasks in communication between a driver and an in-car speech enabled system. This system allows users to accomplish the tasks of making phone calls, listening to the radio, or listening to a music collection. The instances of drivers ending tasks that we analyze below are drawn from driving sessions with twenty-six participants, during which drivers interacted with the in-car speech system. The sessions were on average one and a half to two hours long and took place in western Massachusetts on primarily rural roads. We present excerpts from recordings of these interactions that represent *task endings*. We adopt a cultural discourse theory (Berry 2009; Carbaugh 1988, 2007, 2012; Scollo 2011) perspective in order to identify assumptions about the communication situation that seem to be made by both the participants as well as by the system designers.

Instances demonstrate that drivers used both a voice command, such as "end radio" or "stop music," and the touch screen to end a task. There were also some participants who said "back" or "home" when they wanted to end a task, and many participants who never ended a task at all, but rather switched to another task without ending the first. As the analysis below demonstrates, these task endings suggest that participants are coming to the interaction with expectations regarding the ability of the system to listen continuously to them. Some participants appear to view the communication event with the system as part of a longer communication situation that does not necessarily contain distinct ending points. These assumptions have design implications since there could be adjustments made to accommodate user preference in task ending and user perception of situation boundaries.

THEORETICAL BACKGROUND

Our analysis in this chapter is conducted in the tradition of cultural discourse analysis (CuDA) (Berry 2009; Carbaugh 1988, 2007, 2012; Scollo 2011), stemming from the ethnography of communication (Hymes 1962, 1972, 1974). We treat the car—and interactions within and about it—as a communication situation (Carbaugh et al. 2013, 2014). A communication situation is made up of speech events, which are "directly governed by rules or norms for the use of speech," and these are in turn made up of one or more speech acts, the "minimal term of the set" (Hymes 1974, 52). In CuDA, researchers explore communication situations and events in terms of five key radiants or hubs of meaning active in communication practices, including being, acting, feeling, relating, and dwelling. The focus on these radiants leads to the formulation of cultural propositions, or statements that capture participants' meanings about communication in relation to conceptions of identity, act sequence, emotion, sociality, and location (Carbaugh 2007, 2012; Scollo 2011). In our analysis, we focus on the interactional sequence of communication situations in the car and the meaning of this sequencing for participants who take part in it. We build on past research as part of a larger project exploring cultural characteristics of communication with automotive natural language speech applications (Carbaugh et al. 2012, 2013, 2014; Molina-Markham et al. 2014; Tsimhoni, Winter, and Grost 2009; Winter, Shmueli, and Grost 2013; Winter, Tsimhoni, and Grost 2011).

Previous scholarship has explored the dynamic of how participants end interactions of varying structure and formality in different contexts and cultures and at different points in time (Firth 1972; Goffman 1971, 1974; Hartford and Bardovi-Harlig 1992; Knapp et al. 1973; Schegloff and Sacks 1973; van Over 2014). In achieving closings, participants transition from one communication event or situation to another. In their frequently cited piece on the topic, Schegloff and Sacks (1973) assert that closings are oriented to by conversationalists as having "a proper character" (320); they identify the "closing section" of a conversation as initiated by a "pre-closing" and containing at minimum a terminal exchange, although it may contain other component parts (317). Goffman (1974) notes that closings are connected to participants' framing of an interaction as guided by certain "principles of organization which govern events" (10). Thus, the characteristics of endings depend on what genre of event participants deem a certain encounter to be; for example, bringing to a close an introductory meeting between new colleagues in Finland (Carbaugh 2005) will be done differently than participating in silence at the end of a meeting for business among Quakers (Molina-Markham 2014) or "dissolving" a town meeting in New England (Townsend 2009).

Communication practices of closing are also closely connected to the nature of participants' relationships and the development of those relationships. Goffman (1971) characterizes both greetings and farewells as "access rituals," which he defines as "ritual displays that mark a change in degree of access" (79). How this access is negotiated has important consequences for a relationship. Goffman explains, "the goodbye brings the encounter to an unambiguous close, sums up the consequence of the encounter for the relationship, and bolsters the relationship for the anticipated period of no contact" (79). The importance of leave-taking in reflecting and informing relationships means that, for example, how one takes leave from a social gathering is dependent on the bonds understood to exist between host and guest (Fitch 1990), and the parting of those considered social equals is likely to be different from those who are viewed as having varying degrees of power or control (DuFon 2010). In many cases, strategies for ensuring good relations are employed, such as launching a new topic in a way that demonstrates other attentiveness (Bolden 2008) or blaming an external factor in a pre-closing in order to provide a legitimate reason for why an interaction must end and not imply that one wants to end it (Takami 2002). Bolden notes, "the collaborative nature of leave-taking is an important resource for maintaining and reaffirming interpersonal relationships" (101).

Misunderstanding can occur when participants from different speech communities have different expectations for how an interaction should end. Practices of leave-taking vary widely within and across cultures; for example, Omar (1993) finds that the order of features in Kiswahili closings is less strict than in English closings, and Dogancay (1990) observes that among English speakers, both participants in an interaction may end by saying "goodbye," while among members of a Turkish speech community, there is one phrase used by the person who is leaving, which translates to "I recommend you to God," and a different phrase used in response by the person who is staying, which translates as "laughingly" (60). Scollon and Scollon (1981) explore possible misunderstanding in their examination of Athabaskan-English communication. They explain that while an American or Canadian English speaker may be inclined to end a conversation with an indication that he or she hopes to continue talking with the other person in the future, an Athabaskan, who has been socialized in different communication patterns, would feel it was bad luck to make predictions about the future. Thus, an expression such as "talk to you later," as used by the in-car speech system that we studied here, would be avoided by an Athabaskan. Scollon and Scollon (1981) describe how this avoidance could become problematic, writing, "The Athabaskan, being careful of courting bad luck, may quite unknowingly signal to the English speaker the worst possibility, that there is no hope of getting together again to speak" (27). In this way, the ending of an interaction has consequences for how people view the success of the encounter and the possibility for future positive interactions.

This chapter expands on prior research by examining practices of ending interactions with a new type of interlocutor—an in-car speech activated computer system. While prior research has looked at how people speak with machines (Friedman 1997; Nass and Brave 2005; Nass and Yen 2010; Turkle 2011) and also at how people interact with each other in the car (Laurier, Brown, and Lorimer 2007; Laurier et al. 2008; Haddington 2010), speaking with a computerized speech system in the car has not been as deeply explored in the research literature. In this way, this analysis builds on past understandings of the closing of communication events by looking at a new context. The analysis thus reflects McKay's (2013) notion that "a user interface is essentially a conversation between users and a product to perform tasks that achieve users' goals" (3). McKay asserts that this context is distinctive because the "conversation" is conducted through "the language of UI instead of natural language" (3). This analysis is of use to system designers because it goes beyond describing characteristics of typified users, and instead, focuses on concrete practices of acting in the car. In other words, we highlight concrete ways of enacting a social role through discourse and interaction in a particular cultural context (Geertz 1972). Closings are important to consider in designing speech activated systems because they can influence participants' opinions about the system and willingness to interact further with the system in the future.

We will begin with an overview of the research design, including how driving sessions were organized and how the in-car speech system functioned. We will then turn to an analysis of participants' preferences for ending interactions with the system. This analysis will include a summary of the variety of ways that participants chose to end tasks, as well as a closer look at some instances of task endings in which participant expectations and system actions appear not to have been in alignment. Examination of these instances adds depth and insight into the previously identified preferences. We conclude with a discussion of the implications of these findings, which includes the formulation of a cultural premise that we propose serves as a foundation for, and is informed by, participants' communication practices in this context.

RESEARCH DESIGN

As discussed in previous work (Carbaugh et al. 2012, 2013, 2014; Molina-Markham et al. 2014), this research was conducted in western Massachusetts, and there were a total of twenty-six participants (fourteen female and twelve male). For the study, we equipped participants' own cars with an infotainment interface application with multimodal capabilities. The driver interacted with the interface via a touch screen tablet computer, which we attached to the dashboard of his or her vehicle. Two small speakers broadcast the system's spoken utterances (in a female voice), played the music that the driver requested, and broadcast the recipient's response when a call was placed.

During the drive, one researcher sat in the front seat of the car, and another researcher sat in the rear seat and used a laptop that was connected to the system to respond to the driver's spoken requests. In this way, the researcher in the back seat took over the role of a speech recognizer, but the participant believed that he or she was interacting with a computer system through the touch screen tablet. The entire session was recorded by three small cameras.

At the beginning of the session, participants were introduced to the capabilities of the speech interface, including making phone calls, listening to music, and listening to the radio. Participants were invited to experiment with the system while parked. Following this, they engaged in an off-road test drive in a parking lot, before beginning an on-road driving session. Although participants were encouraged to ask questions during the initial parking lot test, they were instructed to try to ignore the researchers' presence during the on-road session.

About halfway through the on-road driving session, the researcher in the front seat asked the driver to park the car for a brief semi-structured mid-session interview. At this point the researcher would ask the driver about his or her impressions of the system thus far, answer any questions that the driver might have, and introduce the driver to any functions of the system that the driver may not yet have noticed. After the complete on-road driving session, the researcher in the front seat would conduct a second longer in-car interview session. The overall session with each participant lasted approximately one and a half to two hours.

Collaborators working in mainland China have collected data from a similarly designed study. We plan to draw on this data in future work in order to expand our analysis through cross-cultural comparison.

ANALYSIS

Although we initially conducted driving sessions with twenty-six participants, we have chosen not to directly include the first six participants
in the analysis below because the ability to switch between tasks was
not fully operational for these participants, due to system development
delays. (It should be noted that the data from the first six participants
does inform this analysis, however, as indicated by the inclusion of the
excerpt from Participant 5's driving session in the introduction.) Participants initiated a verbal interaction with the system by touching the
microphone button on the touch screen. The system would then produce
a chime sound, or "ding," and the participant would tell the system what
he or she wanted it to do. When they wanted the system to stop doing
something, participants had the option to end the radio or end the music
by either touch or by voice, but they were only able to end phone calls by
touch. Ending a task by touch was done by simply pushing the "end" button on the tablet screen. In order to end a task by voice, it was necessary to
first touch the microphone button and then give a verbal command to end
the task. Phone calls could only be ended by touch because of technological restrictions. During a phone call, it would be difficult for the system
to know whether the driver was talking to the call recipient or giving a
command to the system. There is also the possibility that the driver would
touch the microphone button to give a command during a phone call, but
if the call recipient did not realize that the driver had done this, he or she
might speak and confuse the system because it might interpret what the
call recipient had said as a command. As a result of this possible confusion, the system was designed so that phone calls could only be ended
by touch.

PREFERENCES FOR ENDING TASKS

In reviewing the recordings of the twenty participants, we counted approximately 49 instances in which participants ended a task by voice and
135 instances in which a participant ended a task by touch. One example
of each is given below: Participant 24 ends the music by using a voice
command, and Participant 9 ends a phone call by touching the "end"
button.

Participant 24
31:01

1. ((Music playing))
2. Participant: ((touches microphone button and system dings))

3. System: What kind of music would you like to hear?=
4. Participant: No music. Stop music.
5. ((Music stops playing and system returns to home screen))

Participant 9
24:27

1. Participant: OK, see ya.
2. Phone call recipient: (Bye).
3. Participant: ((Touches end button and system returns to home screen))

Eight of these voice and touch instances overlap, in that the participant used both voice and touch to end the task—such as when Participant 8 said "end" while touching the "end" button on the screen.

Participant 8
24:51

1. Mom: Enjoyed talking to you and uh good luck I hope that th—the results
2. of your experiment continue now that you figured that thing out.
3. Participant: Yup, thanks.
4. Mom: OK. Call me again.
5. Participant: All right. Bye Mom=
6. Mom: Bye. Love you.
7. Participant: End. ((Touches end button as speaks and system returns to home screen)) End call.

These numbers indicate a preference for ending tasks by touch. However, as mentioned, participants were only permitted to end phone calls by touch, which could have inflated the number of touch endings because there was no option in these cases. We found that of the 135 instances in which a participant ended a task by touch, in sixty-four instances, the participant was ending a phone call. In order to account for the requirement to only use a touch command to end a phone call, we could discount these sixty-four cases, and only consider the seventy-one other touch endings (of the music or radio), in which the participant had a choice of touch or voice. It is likely, however, that at least some of the sixty-four phone call endings would have been touch endings even if there had been an option to use voice, so the number of touch endings would likely be greater than seventy-one. Thus, seventy-one (or more) touch endings versus forty-nine voice endings still seems to indicate a preference for ending by touch. The preference for ending by touch may be due to the fact that ending by touch only required one step—that of touching the screen—while ending by voice required two steps: touching the microphone button on the screen and then giving a verbal command.

Although the system had multimodal capabilities, participants were instructed at the beginning of the session to only use voice commands with the system while they were driving—except when they needed to touch the microphone button to start a voice interaction or to touch the end button to end a task. Thus, although participants were permitted to end tasks through touch, one might reasonably expect a pattern of dispreference for this mode to emerge as they had been told not to use touch for all other tasks. However, the high number of touch endings indicates that no such dispreference was active.

We also counted approximately ninety-one instances of participants switching tasks without ending tasks. These switches included changes between radio stations, between listening to the radio or music and making a phone call, and between artists, albums, or songs (we chose not to include going to the next song on an album as a task switch since this switch is not a change to a different task). We have included an instance of a switch between radio stations below. In this case, Participant 20 does not end the sports channel that he is listening to before going to a different radio station.

Participant 20
21:34

1. ((Radio playing))
2. Participant: ((Touches microphone button and system dings))
3. System: What sports channel do you want to hear?
4. Participant: Go to NPR. (5)
5. System: Hold on. (5) Tuning radio to 88.5 FM WFCR.
6. ((Radio station changes))

When compared with the 135 touch endings and forty-nine voice endings, there would seem to be a preference first for touch endings, next for switching tasks, and finally for giving a voice command to end a task. However, if the sixty-four instances of ending phone calls are discounted because they required the participant to use a touch ending, as discussed above, then there is slightly more of a preference for switching tasks (ninety-one) rather than for touch endings (approximately seventy-one). It is, however, difficult to know if participants were simply switching tasks in order to experiment with the system, since they had been told during the introduction to the study that they were helping the researchers to test a prototype. They were also sometimes introduced to the switching functionality during the mid-session interview by the interviewer if they did not discover it on their own during the first part of the drive, which could have increased their likelihood to use this functionality. It is possible that had participants been alone in the car they might not have felt the need to continue to test the system by switching between tasks, but would have simply ended the current task. Therefore, we will now consider in more depth several instances of task endings that appear to have been somewhat problematic for users in order to explore what these instances might reveal about users' task ending practices and preferences.

FURTHER CONSIDERATION OF INSTANCES OF TASK ENDINGS

We will now consider several instances of endings in which there appears to be misalignment between participants' expectations and the system's responses. We will suggest that these misalignments are related to premises of ending interactions.

Touching the Microphone Button

Participants sometimes forgot to touch the microphone button before giving a voice command to end a task, such as Participant 24 below.

Participant 24
23:13

1. ((Radio playing))
2. Participant: End radio. (3) ((Touches end radio button, radio ends and system returns to home screen)) Phone.

Beginning to speak before touching the microphone button occurred frequently throughout the data, not just during task endings. In addition to forgetting to touch the button before ending a task, participants also forgot to touch the button at the beginning of tasks. The example from Participant 24 above was one of at least five instances in which Participant 24 forgot to touch the microphone button before giving a verbal command to either initiate or end a task. Toward the end of her driving session (around fifty-two minutes into the session), Participant 24 forgot to touch the microphone button before giving a verbal command to play the radio, and after this final instance, she reflected aloud:

This reminds me of *Star Trek* when they—I'm a really big *Star Trek: The Next Generation* fan. Sometimes they like hit their communicators to talk, and then sometimes they're just like 'Computer' whatever, like they just talk to like the whole ship, and they're not even like—they're not even hitting their communicator. And it always bothered me cause I always thought, how does the computer know the difference between their conversation in the ship, and like— (laughs) You know they should be like hitting the button every time.

That's what I feel like when I have to keep hitting this button here. I keep forgetting to do it. Cause I keep thinking it's just going to read my mind. (laughs)

Thus, for some participants at least, it seemed natural to simply begin speaking to the system without a need to initiate the interaction by touching a button. Participant 17 noted during the mid-session interview when asked what he thought of the system so far, "It seems kind of strange that you have to keep touching the microphone button. I think it should either be engaged or once you engage it, it should stay that way until you turn it off, I would think." This beginning to speak without initiating the dialog through touch suggests a view of the system as always on or continuously listening—an expectation which is not met by the current system design.

Trouble During Task Switching

Some participants did not seem to realize at first that switching was possible or how to do it. For example, Participant 17 had difficulty figuring out how to accomplish switching. In the instance below he tries to switch XM channels, but since he first says "end radio," the system shuts off before responding to his request for "seven." The system is designed to hear only a statement about desired action, at which point it closes the sequence, while this user's turn design suggests a preference for an ongoing conversation where multiple tasks can be accomplished in sequence moving from ending the current act, to establishing a new act in the same interaction.

Participant 17
1:01:14

1. Participant: ((Touches microphone button and system dings)) End radio.
2. Seven. ((Radio ends and system returns to home screen))

This difficulty in switching happened twice during Participant 17's driving session. As a result, Participant 17 did not switch tasks at all during his session. His attempts at switching became voice endings.

The desire to switch tasks (although unsuccessful) aligns with the tendency to switch that is evident in the ninety-one task switches that other participants did accomplish successfully. Participants' switching reveals an understanding of the communication situation as ongoing, rather than one marked by distinct ending points. Also, since Participant 17's attempts to switch resulted in voice endings, these instances suggest that in some cases when a participant uses voice to end a task, he or she is in fact trying to switch tasks rather than end a task.

Going "Back"

Several participants used the term "back," when they wanted to return to a previous screen and switch tasks. However, the system's response to the command "back" was to repeat the previous action. In the instance below, Participant 23 attempts to use the commands "back" and "home" in order to get to a previous screen before giving a directive for a different task. The system's response of repeating the previous action (lines 4–5) appears to confuse him, as indicated by line 6 when he says, "hmm."

Participant 23
21:09

1. Participant: ((Touches microphone button and system dings))
2. System: What music station do you want to hear?
3. Participant: Um (1.5). Go back.
4. System: Tuning radio to one hundred and four point one FM radio one
5. hundred and four point one WMRQ.
6. Participant: Hmm ((touches microphone button)) (3) Home.
7. System: Could you repeat that please?
8. Participant: Home.
9. System: OK, talk to you later. ((Participant laughs))

Below, Participant 7 attempts to use "go back" to return to the screen that allowed her to choose genres of stations so that she could switch tasks. However, the system responds in line 9 by repeating its previous action from line 3, when it told her that it could not find the artist that she had requested. Participant 7 does not seem to understand that the system has repeated its previous action, and she presses the back button several times in a continued attempt to return to a previous screen.

Participant 7
42:32

1. System: Please review your specified artist once more.
2. Participant: ((unclear))
3. System: I cannot find this artist in your collection. I am sorry.
4. Participant: OK. Um. Let's see. How about, um, something simple. (1.5) Um=
5. System: Pardon?

6. Participant: Mm. (4) Go back to g- Uh, can you go back? (3) Can you go
7. back? ((Touches back button)) Go back.
8. System: I cannot find this artist in your collection. I am [sorry.]
9. Participant: [OK.] OK. OK=
10. System: Please review your specified artist once more.
11. Participant: ((Touches back button again twice)) . . .

On line 5 the system is waiting to hear a recognizable command and apparently hears something on line 8 it understands as the "back" command and takes this action. In doing so, it has returned to the previous user-initiated action, a request to play an artist. Because the participant continues to say "Go back" after the system has re-initiated its prior action, listening for an artist name, "Go back" is likely heard by the system as an artist name that the system cannot recognize. We suggest that the system has interpreted "Go back" in this way based on the system's next turn of notifying the user that this artist cannot be found. At this point, the user responds "OK. OK. OK." perhaps expressing agitation with the interaction, which the system hears as yet another attempt to offer an artist name and responds yet again that this artist cannot be found.

Because the user anticipates that the "back" command will return them to a point further back in the interaction, while the system treats "back" as a return to the user's last turn, a misalignment of expectations for what "back" should accomplish is manifested. This is suggestive of a shift in understanding about when "this" began: a single turn ago, or an event ago. A conflict appears to exist here between machine organization and the organization of conversation between people. While a person interprets interaction in terms of a turn-based sequential organization (Sacks, Schegloff, and Jefferson 1978), a machine functions differently, unless told to do otherwise. A user can input a directive (one turn from the user's view) and the system then responds (another turn from the user's view, making a pair), but from the system's view, in the space of that turn it has taken many micro-actions and executed many internal commands, to bring about the result the user directed. So when a machine "goes back," it goes back to the immediate prior internal command "tuning radio," but when a user says "go back," he or she means go back to the prior pair-part, or two conversational turns.

These examples of different interpretations of the command "back" on the part of the system versus participants again indicate a desire to be able to switch tasks without ending previous tasks. The participant would like to return to an earlier step in the interaction. The interaction is thus seen as part of a larger situation, rather than a distinct and separate event.

DISCUSSION

The frequent use of switching by participants, in addition to certain forms of misalignment in endings, appears to indicate an understanding of the interaction with the in-car system as an ongoing situation that does not begin again completely when the participant touches the microphone button. Misalignments, such as forgetting to press the microphone button before ending a task verbally, trouble switching tasks, and the use of the command "back," reveal a cultural norm that could be formulated as: *when interacting with a system participant (that is, the in-car system), if this is done properly, one should be able to switch to a new task without explicitly ending a previous task.* As Participant 11 explained during the mid-session interview, when the interviewer commented that she had a tendency to go from one task directly to another without ending the first and asked if this practice would be her preference when interacting with the system: "Yeah, I think it should be open and free. Um, so that people don't have to think about, 'Oh, well I want to make a phone call. I got to end the music and then—' Yeah, I think it should flow."

It is important to emphasize that this analysis of communication practices focuses on acting in a specific interactional context rather than on an abstract conception of a particular type of user and what he or she may want generally. It goes further than describing possible personality traits, and instead, draws attention to actions. This analysis suggests that while some users will end interactions with a speech activated system in their car before starting new ones, many participants would prefer to switch between tasks. Unlike when one is interacting with one's smartphone and would perhaps want the phone to "stop listening" altogether at some points—for example, when one places the phone in one's pocket or purse—the preference for switching in the car indicates that perhaps participants do not want their in-car system to stop listening. In other words, participants do not want a change in "degree of access," as Goffman (1971) would describe it (79).

The reasons behind this different preference would need to be explored further. There may be different conceptual boundaries at work here, with system designers employing what we might call an event-based boundary in which a variety of clearly delineated events occur in any given drive, while participants experience a situational boundary, where entering and exiting the car-as-situation signifies the "beginning" and "end." There is also perhaps a parallel to what Schegloff and Sacks (1973) describe as the "continuing state of incipient talk" that occurs between human co-passengers in a car, who can be silent for a period of time without closing an interaction or needing to "begin new segments of conversation with exchanges of greetings" (325). In addition, participants may have preferred switching because in one's car, one's primary focus is driving. Consequently, minimal interaction with an in-car speech system, such as that required by switching rather than ending interactions, would likely be more desirable. This preference for fewer steps and, thus, potentially less involved interaction is also seen in the tendency to use touch versus voice endings, since voice endings required two steps, while touch endings required only one. Participant 9 observes that overall in interacting with the system he feels that "It's easier to touch it, so it's like if I wasn't driving, that would be my choice." However, he *is* driving and his comment emphasizes that the in-car situation is a complex environment in which one is balancing different activities that require varying degrees of attention and visual engagement at different points, making preferences for interaction perhaps more fluid and dynamic than other contexts, as one switches between driving down the highway or waiting at a stoplight, for example. Murray (2012) observes that every new design medium (for example, the emerging digital medium of an in-car speech enabled system) offers unique affordances, and the design conventions of past media must be changed or adapted to take advantage of these new affordances. Therefore, the conventions for designing past digital media, such as tablets or smartphones, might not be appropriate in designing an in-car system, especially in terms of situational boundaries or demands on visual attention.

An important issue raised by this analysis is the extent to which practices of ending or switching tasks in interactions with an in-car voice activated system differ between cultures. In particular, there may be different conceptual boundaries informing interaction in different cultures. We are currently working on extending our analysis to include a comparison with interactions occurring with an in-car system in mainland China. At present, we have only examined data from a pilot study of three English-speaking Chinese drivers in mainland China interacting with the system in English, rather than Chinese.

Chinese participants in the pilot study would sometimes try to end a task with a spoken directive and then switch to touching the "end" button when the system did not respond right away. Chinese participants also used a wider variety of English words for verbal endings (for example, "end," "exit," "abort," "forget") than American participants. Attempts to give verbal directives by Chinese participants could be explained by the fact that they were in an English-speaking situation. Participants might have felt obliged to try out the system in English regardless of a possible preference for one modality (that is, touch) over the other (that is, voice). Further analysis is needed to determine the extent to which Chinese participants practiced switching tasks and what this might indicate about how situational boundaries inform interaction.

CONCLUSION

Researchers suggest that user expectations and preferences when interacting with speech enabled technology differ culturally. System designers make assumptions about the users who will interact with their system and must be careful to take into account different cultural premises of communication. Adopting a cultural discourse theory perspective, we analyzed communication practices of task endings in communication between drivers in the northeastern United States and an in-car speech enabled system. Our analysis reveals that while participants ended tasks through touch and voice, they also frequently chose to simply switch tasks rather than explicitly end them. By exploring instances of task endings in which user and system designer expectations appear to have differed, we identified a possible cultural norm at play in the communication situation of the car. We suggest that the expected boundary of the communication situation for users is different from the boundary used in the initial design choices of system designers. It appears that task endings in this context of human-machine interaction in an automobile differ from other contexts of closings between people and between people and other forms of voice activated technology.

Cultural premises about the boundary of the interaction may also inform other premises about sociality and personhood. If the system is conceived as a conglomerate of discrete events, then perhaps one's relationship with that system is also discrete, manifesting and diffusing with the start and end of each event. This in turn may connect to premises about personhood, in which the car is imagined as an instrumental machine to be turned on or off in order to accomplish tasks and nothing more. If, however, the boundaries of the system are situational, and not event-based, then the system is always there, and the relationship continues during and beyond the verbal interaction one has with it, more akin to human-human relationships. By extension, premises of personhood may shift with this boundary; the system is conceived not as an instrument that goes away when it is done completing a task, but as an interactional partner, waiting and listening until you leave. Of course these premises may not be related in this way in practice—although other research does suggest that people may bring different premises about personhood to their in-car interactions (Molina-Markham et al. 2014). This is meant only to demonstrate possible connections between cultural conceptions of the boundaries of an interaction and the social relations and models of personhood they implicate. This type of research can lead to improvements or enhancements that are constantly part of the iterative design and redesign process. Thus, it is important to adopt an approach of incorporating findings from research on participant interaction with technology during initial design stages.

This work was supported by General Motors and Donal Carbaugh, principal investigator, in collaboration with Ute Winter, research specialist of the Human-Machine Interface at General Motors Research and Development, Herzeliya, Israel; Elizabeth Molina-Markham, post-doctoral research fellow; and Brian van Over and Sunny Lie, research associates.

REFERENCES

Berry, Michael. "The Social and Cultural Realization of Diversity: An Interview with Donal Carbaugh." *Language and Intercultural Communication* 9, no. 4 (2009): 230–241.

Bolden, Galina. "Reopening Russian Conversations: The Discourse Particle *-to* and the Negotiation of Interpersonal Accountability in Closings." *Human Communication Research* 34 (2008): 99–136.

Carbaugh, Donal. *Talking American: Cultural Discourses on Donahue*. Norwood, NJ: Ablex, 1988a.

Carbaugh, Donal. *Cultures in Conversation*. Mahwah, NJ: Lawrence Erlbaum Associates, Inc., 2005.

Carbaugh, Donal. "Cultural Discourse Analysis: Communication Practices and Intercultural Encounters." *Journal of Intercultural Communication Research* 36, no. 3 (2007): 167–182. doi:10.1080/17475750701737090.

Carbaugh, Donal. "A Communication Theory of Culture." In *Inter/Cultural Communication: Representation and Construction of Culture*, edited by Anastacia Kurylo, 69–87. Thousand Oaks: Sage, 2012.

Carbaugh, Donal, Elizabeth Molina-Markham, Brion van Over, and Ute Winter. "Using Communication Research for Cultural Variability in Human Factor Design." In *Advances in Human Aspects of Road and Rail Transportation*, edited by Neville Stanton, 176–185. Boca Raton, FL: CRC Press, 2012.

Carbaugh, Donal, Ute Winter, Elizabeth Molina-Markham, Brion van Over, Sunny Lie, and Timothy Grost. "A Model for Investigating Cultural Dimensions of Communication in the Car." Manuscript submitted for publication, 2014.

Carbaugh, Donal, Ute Winter, Brion van Over, Elizabeth Molina-Markham, and Sunny Lie. "Cultural Analyses of In-Car Communication." *Journal of Applied Communication Research* 41, no. 2 (2013): 195–201.

Coupland, Nikolas, and Justine Coupland. "Language, Ageing and Ageism." In *The New Handbook of Language and Social Psychology*, edited by Peter Robinson and Howard Giles, 465–486. New York: John Wiley, 2001.

Dogancay, Seran. "Your Eye is Sparkling: Formulaic Expressions and Routines in Turkish." *Working Papers in Educational Linguistics* 6, no. 2 (1990): 49–64.

DuFon, Margaret. "The Socialization of Leave-Taking in Indonesian." *Pragmatics and Language Learning* 12 (2010): 91–111.

Firth, Raymond. "Verbal and Bodily Rituals of Greeting and Parting." In *The Interpretation of Ritual,* edited by J. S. La Fontaine, 1–38. London: Tavistock, 1972.

Fitch, Kristine. "A Ritual for Attempting Leave-Taking in Colombia." *Research on Language & Social Interaction* 24, no. 4 (1990): 209–224.

Friedman, Batya, ed. *Human Values and the Design of Computer Technology.* New York: Cambridge University Press, 1997.

Geertz, Clifford. "Deep Play: Notes on the Balinese Cockfight." *Daedalus* 101, no. 1 (1972): 1–37.

Goffman, Erving. *Relations in Public: Microstudies of the Social Order.* Philadelphia, PA: Harper and Row, 1971.

Goffman, Erving. *Frame Analysis: An Essay on the Organization of Experience.* Boston, MA: Northeastern University Press, 1974.

Haddington, Pentti. "Turn-Taking for Turntaking: Mobility, Time, and Action in the Sequential Organization of Junction Negotiations in Cars." *Research on Language and Social Interaction* 43, no. 4 (2010): 372–400.

Hartford, Beverly S., and Kathleen Bardovi-Harlig. "Closing the Conversation: Evidence from the Academic Advising Session." *Discourse Processes* 15, no. 1 (1992): 93–116.

Hymes, Dell. "The Ethnography of Speaking." In *Anthropology and Human Behavior,* edited by Thomas Gladwin and William Sturtevant, 13–53. Washington, DC: The Anthropological Society of Washington, 1962.

Hymes, Dell. "Models for the Interaction of Language and Social Life." In *Directions in Sociolinguistics: The Ethnography of Communication,* edited by John Gumperz and Dell Hymes, 35–71. New York: Basil Blackwell Inc., 1972a.

Hymes, Dell. *Foundations in Sociolinguistics: An Ethnographic Approach.* Philadelphia, PA: University of Pennsylvania Press, 1974.

Knapp, Mark L., Roderick P. Hart, Gustav W. Friedrich, and Gary M. Shulman. "The Rhetoric of Goodbye: Verbal and Nonverbal Correlates of Human Leave-Taking." *Communication Monographs* 40, no. 3 (1973): 182–198.

Laurier, Eric, Barry Brown, and Hayden Lorimer. *Habitable Cars: The Organisation of Collective Private Transport.* Full Research Report to the Economic Social Research Council (University of Edinburgh, University of Glasgow), RES-000–23–0758. Swindon: ESRC, 2007.

Laurier, Eric, Hayden Lorimer, Barry Brown, Owain Jones, Oskar Juhlin, Allyson Noble, Mark Perry, Daniele Pica, Philippe Sormani, Ignaz Strebel, Laurel Swan, Alex S. Taylor, Laura Watts, and Alexandra Weilenmann. "Driving and 'Passengering': Notes on the Ordinary Organization of Car Travel." *Mobilities* 3, no. 1 (2008): 1–23.

McKay, Everett N. *UI is Communication: How to Design Intuitive, User Centered Interfaces by Focusing on Effective Communication.* Waltham, MA: Elsevier, 2013.

Molina-Markham, Elizabeth. "Finding the 'Sense of the Meeting': Decision Making Through Silence among Quakers." *Western Journal of Communication* 78, no. 2 (2014): 155–174.

Molina-Markham, Elizabeth, Brion van Over, Sunny Lie, and Donal Carbaugh. "'You Can Do It Baby': Non-Task Talk with an In-Car Speech Enabled System." Manuscript submitted for publication, 2014.

Murray, Janet H. *Inventing the Medium: Principles of Interaction Design as a Cultural Practice*. Cambridge, MA: The MIT Press, 2012.

Nass, Clifford Ivar, and Scott Brave. *Wired for Speech: How Voice Activates and Advances the Human-Computer Relationship*. Cambridge, MA: MIT Press, 2005.

Nass, Clifford Ivar, and Corina Yen. *The Man Who Lied to His Laptop: What Machines Teach Us about Human Relationships*. New York: Current, 2010.

Omar, Alwiya S. "Closing Kiswahili Conversations: The Performance of Native and Non-Native Speakers." *Pragmatics and Language Learning* 4 (1993): 104–125.

Philipsen, Gerry. *Speaking Culturally: Explorations in Social Communication*. Albany, NY: State University of New York Press, 1992.

Sacks, Harvey, Emanuel Schegloff, and Gail Jefferson. "A Simplest Systematics for the Organization of Turn-Taking for Conversation." In *Studies in the Organization of Conversational Interaction*, edited by Jim Schenkein, 7–55. New York: Academic Press, 1978.

Schegloff, Emanuel, and Harvey Sacks. "Opening Up Closings." *Semiotica* 8, no. 4 (1973): 289–327.

Scollo, Michelle. "Cultural Approaches to Discourse Analysis: A Theoretical and Methodological Conversation with Special Focus on Donal Carbaugh's Cultural Discourse Theory." *Journal of Multicultural Discourses* 6, no. 1 (2011): 1–32.

Scollon, Ronald, and Suzanne Scollon. *Narrative, Literacy and Face in Interethnic Communication* 7. Norwood, NJ: Ablex Publishing Corporation, 1981.

Takami, Tomoko. "A Study on Closing Sections of Japanese Telephone Conversations." *Working Papers in Educational Linguistics* 18, no. 1 (2002): 67–85.

Townsend, Rebecca. "Town Meeting as a Communication Event: Democracy's Act Sequence." *Research on Language and Social Interaction* 42, no. 1 (2009): 68–89.

Tsimhoni, Omer, Ute Winter, and Timothy Grost. "Cultural Considerations for the Design of Automotive Speech Applications." Paper presented at the 17th World Congress on Ergonomics IEA, Beijing, China, 2009.

Turkle, Sherry. *Alone Together: Why We Expect More from Technology and Less from Each Other*. New York: Basic Books, 2011.

van Over, Brion. "Tracing the Decay of a Communication Event: The Case of *The Daily Show's* 'Seat of Heat.'" *Text & Talk* 34, no. 2 (2014): 187–208.

Winter, Ute, Yael Shmueli, and Timothy Grost. "Interaction Styles in Use of Automotive Interfaces." In *Proceedings of the Afeka AVIOS 2013 Speech Processing Conference*, Tel Aviv, Israel, 2013.

Winter, Ute, Omer Tsimhoni, and Timothy Grost. "Identifying Cultural Aspects in Use of In-Vehicle Speech Applications." In *Proceedings of the Afeka AVIOS 2011 Speech Processing Conference*, Tel-Aviv, Israel: Literature, 2011.

Two

Analyzing Procedure to Make Sense of Users' (Inter)actions

A Case Study on Applying the Ethnography of Communication for Interaction Design Purposes

Tabitha Hart[1]

The information communication technologies of our early twenty-first century support an astonishingly complex range of person-to-person interactions, from the local to the global, from mundane to extraordinary, for purposes modest to lofty. Designing user experiences for today's global, technology-mediated interactions is no simple matter, particularly when platforms are intended to connect people across linguistic and cultural borders, via a multiplicity of channels and modes. What's more, such platforms must often serve different purposes for multiple stakeholders, such as whole organizations, their service providers, and their clients/users. Utilizing a local strategies research perspective can be helpful in navigating this multifaceted design terrain. In this chapter, I describe two related conceptual tools, act sequence and procedural knowledge, which are grounded in the ethnography of communication research tradition. Using a case study on Eloqi,[2] a virtual organization that built and deployed an online English as a foreign language (EFL) training program for paying customers in China, I will demonstrate how act sequence and procedural knowledge can be used to examine local understandings about acting, action, and practice in technology-mediated settings. More specifically, I will use these key concepts to analyze problematic user experiences that occurred during live interactions between Eloqi's employees (English trainers) and their clientele (students). To situate my study I discuss the theoretical context for this work, introducing pertinent concepts drawn from the ethnography of communication and outlining their relevance to interaction design. I then present the research context for this case study, followed by the data analysis and findings. Finally, I suggest the broader implications of this research.

THEORETICAL CONTEXT

The ethnography of communication (EC) is a distinct theoretical and methodological approach to studying situated communication practices as well as the local cultures and strategies that such practices instantiate. EC is closely related to ethnography, a social scientific research tradition rooted in the discipline of anthropology. Like ethnography, which "[discerns] patterns of socially shared behavior" (Wolcott 1999, 67), EC research is used to produce ethnographic reports detailing and interpreting local cultural processes. As with traditional ethnography, EC typically involves immersion in a local setting, during which time the researcher employs various methods of data collection, primarily qualitative (participant observation, interviews, etc.) but possibly quantitative, too. EC is differentiated from ethnography by its lineage and focus: it was born from linguistics, focuses on communication practices, and uncovers "relationships between language and culture" (Keating 2001, 285). More specifically, by examining the patterning of communication norms, rules, practices, and meanings, EC-grounded research can effectively discern local beliefs about personhood (what it means to be a person in the world), sociality (how to connect with others in a community), and rhetoric (how to communicate strategically to achieve one's desired goals) (Philipsen and Coutu 2005; Carbaugh 2005, 2007; Philipsen 2002; Philipsen, Coutu, and Covarrubias 2005).

In the last twenty-five years, EC scholars have produced substantial reports analyzing the communication practices and traditions of local communities. This body of work represents a wide variety of languages and cultures, and includes both intercultural analyses as well as cross-cultural comparisons (Baxter 1993; Carbaugh 1988, 2005; Coutu 2000; Edgerly 2011; Katriel 1986; Katriel and Philipsen 1981; Philipsen 1975, 1992, 2000; Philipsen and Leighter 2007; Winchatz 2001; Fong 2000; Leighter and Black 2010; Sprain and Gastil 2013; Witteborn and Sprain 2009). There is now a growing interest in using EC-grounded approaches to study online and other technology-mediated communication, whether to examine the communication and cultural life of online communities or the ways in which people interact with technologies offline (Carbaugh et al. 2013; Dori-Hacohen and Shavit 2013; Witteborn 2011, 2012; Boromisza-Habashi and Parks 2014; Hart 2011). Just as communication scholarship in general can contribute to design work (Jackson and Aakhus 2014; Aakhus and Jackson 2005), EC has much to offer towards the strategic design of communication structures, actions, and practices (Leighter, Rudnick, and Edmonds 2013; Sprain and Boromisza-Habashi 2013), including those for technology-mediated environments. In fact, several key characteristics of EC research make it a good fit for user experience research and interaction design.

User interfaces (UIs) are a means not only of presenting information, options, and activities to the user, but also of organizing information, options, and activities. As such they are communication tools that support communication processes, and they embody, employ, strategically utilize, and support communication conventions. In basic design terms, the UI must communicate its functionalities and protocols to the users clearly enough that they can easily intuit what to do, when, and how (McKay 2013). The UI, however, is just one component of the larger user experience (UX), which "encompasses the entire experience users have with a product [including] the internals that users don't interact with directly, as well as the externals, such as the purchasing process, the initial product experience (often called the 'out-of-box' experience), customer and technical support, product branding, and so on" (McKay 2013, 6–7). The ultimate aim in designing a UX for a technology-mediated environment is to foster the possibility for what is dubbed agency.

> Agency results when the interactor's expectations are aroused by the design of the environment, causing them to act in a way that results in an appropriate response by the well-designed computational system. This matching of the interactor's participatory expectations and the actions to the procedural scriptings of the machine creates the pleasurable experience of agency. Bad design frustrates the interactor by creating confusing or unsatisfiable expectations, or by failing to anticipate actions by scripting the machine with appropriate responses. (Murray 2012, 12–13)

In other words, the ideal technology-mediated environment invites instinctive actions that match users' own "mind maps" for engaging in the task and/or interaction at hand. If the user can act instinctively in the environment and produce the appropriate (anticipated, desired) results, then the design is a success.

Achieving the desired degree of agency in a build may be complicated by the fact that the build itself (the UI, or the technology supporting the communication) shapes the process of using it (Appel et al. 2012), sometimes in unexpected ways. Presumably, designing for maximum agency becomes even more complex when the build connects users for person-person interactions, whether asynchronous or synchronous, or via text, audio, and/or video. In these cases the design has an immediate impact not only on the user-machine interaction, but also on the user-user interactions being supported by the technology (Appel et al. 2012). In these cases, designers must account for multilayered and complex sociocultural dynamics impacting the user experience: users' social orientations towards their interactions with the technology (Nass, Steuer, and Tauber 1994), "the interpretation of [technological] artifacts as part of larger social and cultural systems" (Murray 2012, 11), the interactions of users with other users via the technology (Dix et al. 2004), and the sometimes nebulous social conventions that users develop for use in particular technology-mediated communication situations (Vorvoreanu 2009). Because of this, it makes sense to include sociocultural analysis into UX design, all the better to understand how "design decisions that shape [digital artifacts] affect the way we think, act, understand the world, and communicate with one another" (Murray 2012, 2).

The definition of agency presented above, particularly as it applies to user experience and design, strongly parallels a concept central to EC: communicative competence. Communicative competence is the ability to communicate appropriately with others according to the local norms, premises, rules, and other socio-linguistic factors of the given context (Hymes 1972a, 1972b; cf. Sprain and Boromisza-Habashi 2013; Witteborn 2003). From the EC perspective, standards of communicative competence are applied in all social groups, across all potential means, modes, and styles of communication. What those standards of communicative competence are, however, will vary widely according to the local setting, participants, goals, norms, etc. (i.e. the local culture). For this reason, defining communicative competence always necessitates carefully identifying how one is expected to communicate properly according to the local culture and the given circumstances (Philipsen 2010). As the above definition of agency suggests, this is precisely the aim of good user design. To produce good builds, designers must thus be highly attentive to the social conventions (norms, premises, rules, etc.) associated with technology use. These conventions include those "that govern our navigation of space, our use of tools, and our engagement with media" (Murray 2012, 10) as well as those governing users' interactions with one another. As sociocultural artifacts, some of these conventions may be universal (culturally general), but they are likely to include local (culturally specific) conventions, too. Whether designing a communication tool or a strategic communication process, the objective is to create a build that fits with and leverages users' intuitive, locally endorsed ways of being, connecting, and communicating (Leighter, Rudnick, and Edmonds 2013; Sprain and Boromisza-Habashi 2013). The EC approach provides us a means of discovering these locally endorsed ways (Hymes 1962, 1972a; Saville-Troike 1982).

Being communicatively competent requires acting in accordance with context-specific variables (Philipsen 2000) such as the setting, participants, goals, norms, etc. These variables are neatly summarized in the SPEAKING heuristic (Milburn n.d.; Hymes 1962, 1972a), an EC tool for analyzing situated communication summarized in Table 2.1. Here I call out one variable in particular: the act sequence. Act sequence denotes the sequence, or order, in which a communicative activity is expected to play out (Hymes 1962, 1972a; Saville-Troike 2003). Act sequences for everyday

Table 2.1. Hymes's SPEAKING Heuristic

Setting/Scene	What is the setting in which the communication activity is taking place?
Participants	Who is involved in the communication activity? What are their roles and relationships?
Ends	What are the goals of this communication activity?
Act sequence	What are the activities comprising the communication activity, and how are they sequenced?
Key	What is the tone of the communication/activity?
Instrumentalities	How is the communication being carried out? Through what modes and/or means?
Norms	What are the social norms governing communication here?
Genre	What is the genre or style of this communication activity?

Table created by Tabitha Hart referencing work by Dell Hymes. Please see this chapter's references for a complete list of Hymes's works utilized.

and routinized behaviors are "conventionalized" patterns of communicative behavior, often distinct to the local cultural milieu in patterning, form, and/or content (Hymes 1962, 1972a; Saville-Troike 2003). We naturally draw on our learned, localized understandings of act sequences as we engage in tasks, social situations, and other types of routine activities, including those mediated by technology. With, for example, a work-related email, the standard act sequence would be a salutation followed by the main point of the message, with a valediction at the close.

To know an act sequence for a given activity is equivalent to possessing procedural knowledge, that is, the knowledge of what steps or actions should occur, how they should be carried out, and in what particular sequence (Shoemaker 1996; Nickols 2000). Here again, there is a clear connection between UX and EC: good design leverages users' procedural knowledge and engages users in act sequences that feel natural and logical. Where "a poorly designed UI is unnatural…and requires users to apply thought, experimentation, memorization, and training to translate it into something meaningful" (McKay 2013, 3; cf. Nielsen 1994; Nielsen 2015), a good design presents users with a natural "fit" between the procedural knowledge that they hold in mind and the act sequencing built into the design. Importantly, the EC approach provides a theoretical/methodological approach to identifying what act sequences are considered natural or logical in local contexts, thereby aiding in the process of inventorying users' procedural knowledge. It can be challenging to articulate procedures, given the innateness of this type of knowledge,[3] so this is a very useful feature of EC.

An opportune situation for identifying procedure is clash, or cases in which interlocutors apply different and/or conflicting notions of procedural knowledge (Shoemaker 1996; Holdford 2006; Bailey 1997). The EC approach is especially well suited to studying such cases and helps researchers attune to "the differences in communication practices that lie at the root of different social, technical, or environmental disputes or miscommunication" (Sprain and Boromisza-Habashi 2013, 183). Numerous EC reports have been produced that identify and examine cases of communication tension and clash in real life settings (Coutu 2000, 2008; Bailey 1997; Huspek 1994).

Finally, once local concepts of communicative competence and proper act sequence have been identified, EC findings can be used to "suggest modes of intervention that resonate with local needs and local systems of meaning" (Sprain and Boromisza-Habashi 2013, 182; Sprain and Gastil 2013), making it a perfect fit for the iterative design/redesign approach favored in the field of UX (Cooper 2004).

To summarize, the EC approach is tailor-made for focusing on real users rather than imagined ones, actual practices rather than assumed ones, and local concepts of natural and correct communication as performed and described by users themselves. In all of these senses, EC research is truly a user-centered approach (Witteborn 2012; Witteborn, Milburn, and Ho 2013) and highly suited to UX/design purposes.

RESEARCH CONTEXT

Eloqi (2006–2011) was a small startup company that built and deployed an online English as a foreign language (EFL) training program for paying customers across China. Eloqi's training program focused on oral communication skills and was designed to help customers pass the oral component of the IELTS, an internationally recognized English proficiency exam. By logging into Eloqi's password-protected spaces, customers could access the company's specialized learning modules (lessons, homework assignments). More importantly, they could use the company's interactive, web-based, and voice-enabled UI to connect one-to-one with English trainers in the United States for live fifteen-minute conversation lessons.

With Eloqi's express support I conducted an ethnographic study of the company, whose members (students, trainers, and admins) met almost entirely online. The most important period of my study was the ten months from 2009 to 2010 when I conducted online participation observations within the Eloqi community. As a participant observer I was inducted into the Eloqi trainer pool. In this role I participated regularly in the community's online activities, reading and responding to posts in the trainer forum, attending weekly trainer meetings, working shifts, and hanging out with the other trainers—all online. Most importantly, I worked directly with Eloqi's students, training them in English conversation skills in intensive one-to-one fifteen-minute sessions, just as the other trainers did.

At the time of my participant observations, Eloqi's most popular lessons were those in the Core English Logic (CEL) series, which the company developed expressly to prepare students for the oral component of the IELTS. The CEL series was the brainchild of the company's chief technology officer, who had assembled a team to crack the code of the IELTS oral exam. After researching the types of questions posed to candidates, this team identified what they believed to be a comprehensive set of thirty-one common IELTS question formulations. Accordingly, Eloqi created the CEL lesson series to teach students clear-cut strategies for classifying and answering each of these questions types, a sampling of which is presented in Table 2.2.

To access the CEL lesson series, students contacted the Eloqi office (located in Beijing, China) by phone or email to purchase a subscription. Once subscribed, the students were free to access the Eloqi platform, where they could choose which lessons they wanted to do during the available timeslots of their choice. Once a student had initiated a lesson, he or she would use an Internet-enabled device to work through a self-guided online pre-activity. All pre-activities were designed to prepare students for their live interactions with trainers, and included materials on relevant vocabulary, pronunciation, grammar, and so on. After completing the pre-activity, students would be placed in an online queue to be connected with the next available trainer. When the student's turn

Table 2.2. Eloqi CEL Question Types and Recommended Answer Strategies

CEL Question Type	Eloqi's Recommended Strategy for Answering
How often do you do X?	To talk about how often you do something, state how often you do it. Explain why you do it at that particular degree of frequency. Give detailed reasons. Give examples.
What do you usually do?	To speak about what you usually do, state what you do when you get up in the morning. Next, state what you do at different parts of the day. Say how often you do these things (sometimes, never, frequently). Finally, say how you feel about them.
What do you dislike about X?	First you say one or two things that you don't like, say how much or the degree that you don't like it, and say why you don't like it.

Table created by Tabitha Hart using Eloqi lesson materials. Published with the knowledge of the company identified by the pseudonym Eloqi.

came up, the system would automatically connect her/him with an Eloqi trainer. Together, the trainer and student would follow the prompts on their screens to proceed through the lesson that the student had chosen.

Each CEL lesson was structured around a fixed sequence of increasingly complex tasks and activities to teach the given formula. Because Eloqi desired a high degree of control over and consistency in the use of its proprietary learning materials, the company scripted all CEL lessons and also built the scripting into the UI. A typical CEL lesson opened with a very brief greeting before proceeding directly to pronunciation practice with the target vocabulary. This was followed by a series of short drills during which the student practiced building phrases and statements that could be used to answer the relevant CEL question type. Finally, the lesson transitioned into a "putting it all together" segment, during which the student practiced answering the target question in a slightly more conversational manner. For each of these CEL lesson segments, the UI presented the trainer with prompts on what to say and when to say it (Figure 2.1, a). While some of the prompts in the UI were open enough to allow trainers to select their own phrasing ("correct [the student]," "reformulate [the question]," "ensure the student understands"), many were fully scripted ("Now let's practice answering the questions like in a real exam; your answers should last for forty seconds at the most") and were intended to be read out word-for-word. When trainers recited lines or successfully led a student through a section, they clicked the corresponding prompt in the UI, causing the prompt and the section to "white out," denoting comple-

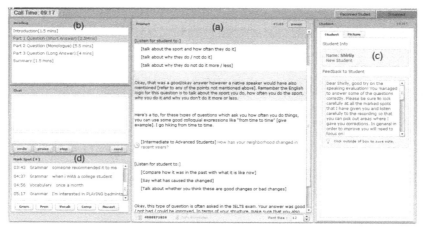

Figure 2.1.   Eloqi lesson UI. Screen shot published with the knowledge and agreement of the company identified by the pseudonym Eloqi.

tion. This action was always recorded in the system and was visible to admins as well as any other trainers who subsequently worked with that student on that lesson. In this way the UI served as a visual tracking cue by which a viewer could quickly see evidence of how the trainer and student had progressed through the lesson.

Simultaneous to working through the prompts and the highly structured lesson plan, trainers had to carefully manage their time. Each segment had its own time limit (Figure 2.1, b) and the entire lesson could not run more than fifteen minutes. What's more, trainers were required to give a minimum number of corrections to each student, both orally and in writing. Personalized notes to each student were provided in a dedicated feedback box on the UI (Figure 2.1, c), while detailed written corrections to the student's speech were provided in a separate area of the screen (Figure 2.1, d).

During my participant observations I jotted down notes and took screen shots, and after each observation I wrote up field notes (Emerson, Fretz, and Shaw 1995). In addition to recordings of my own lessons, Eloqi also granted me unrestricted access to the company's master archive, which contained audio files and screen shots documenting every trainer-student interaction that occurred on the platform. From this archive I selectively transcribed and analyzed recordings that were relevant to the experiences, discussions, and activities of trainers and students. Ultimately I reviewed approximately 130 trainer-student recordings and transcribed about half of them. Finally, I conducted a series of interviews with Eloqi admins, trainers, and students. The aim of these interviews was to investigate points of interest that arose during my participant observation and ask interviewees about their perceptions and interpretations of the Eloqi experience. All of this material (notes, screen shots, field notes, trainer-student lesson transcriptions, interview transcriptions) became part of my dataset.

In preparing the original write-up for this study my goal was to identify the system of norms, premises, and rules guiding communicative conduct, that is, the *speech code* (Philipsen 1997; Philipsen, Coutu, and Covarrubias 2005), in Eloqi's community. This included the community's key values on personhood (what it meant to be a competent English speaker), relationships (how trainers and students connected with one another on a relational/interpersonal level), and rhetoric (what it meant to communicate with one another strategically), which I have reported on elsewhere (Hart 2016).

Meanwhile, as I was collecting and analyzing the data I discovered an intriguing subset of trainer-student interactions in which the lessons did not go as planned. I approached these interactions as speech events, or routinized speech activities "directly governed by rules or norms for the use of speech" (Hymes 1972a, 56; cf. Saville-Troike 2003). From there, I applied Hymes's SPEAKING heuristic (Hymes 1962, 1972a) to analyze which, if any, of these communicative competence-related variables helped to explain what was happening. In so doing, I found the act sequence variable combined with the related concept of procedural knowledge to be very helpful in making sense of what was not working in these lessons. I now turn to an analysis of what this process yielded and an explication of how the concepts of act sequence and procedural knowledge shed light on why these interactions were problematic.

ANALYSIS

My discovery of these cases of problematic communication occurred in one of four ways: a flag in the system marked the case as problematic; a colleague reported issues to the community; I experienced the issues myself while teaching; or I came across a case while transcribing and analyzing trainer-student recordings. Most of the cases of problematic communication that I examined were associated with, or resulted in, the following conditions:

1. Early termination of an interaction by a trainer or a student. Each trainer-student interaction was required to run a minimum of twelve minutes. If a lesson ran significantly under this minimum, it was red flagged in the system as incomplete.
2. Directives by a trainer to a student to call HST (Eloqi's customer service team) for assistance. HST representatives were charged with interfacing directly with students to solve any problems that arose.
3. Reports by trainers to supervisors about problematic communication with a student. All trainers were required to "hang out" in the trainer chat room (Figure 2.2) while working. Beyond being a convivial space for passing the time in between lessons, the chat room was where trainers reported any issues with students. Whenever issues arose, trainers announced them in the chat room. The supervisor on duty in the chat room would then contact HST, and HST would in turn contact the student to bring the issue to resolution.
4. Technical issues that slowed or halted a lesson, or caused it to terminate, including audio/sound problems, the UI not responding properly, and other difficulties related to the technological aspects of the platform.
5. Markedly halted progress through a lesson. As previously mentioned, lessons were strictly timed, and the total lesson time had to fall between twelve and fifteen minutes.[4] Each lesson was comprised

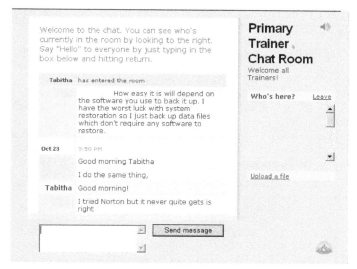

Welcome to the chat. You can see who's
currently in the room by looking to the right.
Say "Hello" to everyone by just typing in the
box below and hitting return.

Primary
Trainer
Chat Room
Welcome all
Trainers!

Tabitha has entered the room

How easy it is will depend on
the software you use to back it up. I
have the worst luck with system
restoration so I just back up data files
which don't require any software to
restore.

Who's here?   Leave

Oct 23   9:50 PM

Good morning Tabitha

I do the same thing,

Upload a file

Tabitha   Good morning!

I tried Norton but it never quite gets is
right

Send message

**Figure 2.2. Eloqi trainer chat room. Screen shot published with
the knowledge and agreement of the company identified by the
pseudonym Eloqi.**

of a series of tasks and activities, and each of these in turn had an allotted number of minutes, meaning that the trainers had to maintain
a pre-determined pace throughout the interaction. When I observed
that a trainer was spending significantly longer than the allotted
time on a given activity, I categorized it as markedly halted progress.
Occasionally trainers reported this in the chat room.
6. Significant deviations from the standard Eloqi lesson script. As
   described earlier, all Eloqi lessons were heavily scripted and pre-
   planned. When I observed that a trainer-student interaction was
   straying from the lesson script in a significant and/or sustained
   manner, I categorized it as a script deviation.

In analyzing these cases, I found that the vast majority of them pertained to misunderstandings around the expected act sequence for
trainer-student interactions. In other words, trainers and students experienced confusion about how to competently proceed through the lesson
according to the local Eloqi lesson protocols. What's more, these cases
could be sorted into four broad types of procedures, summarized in Table
2.3, each of which I will now describe.

Table 2.3. Procedure Types

Procedure Type	Summary
Lesson Initiation & Participation	How to initiate and participate in an Eloqi lesson; how to meet the expected conditions for participation.
Navigation/UI	How to navigate and use features of the Eloqi UI within the context of a live lesson with an Eloqi trainer.
Task/Activity Content	How to complete specialized CEL tasks and activities, as per the task/activity design.
Troubleshooting the Technology	How to handle technical problems that arise during a live Eloqi lesson.

Data collection and table creation by Tabitha Hart.

LESSON INITIATION AND PARTICIPATION: HOW TO TAKE PART IN AN ELOQI LESSON

The most fundamental requirement for participating in an Eloqi lesson was to be seated at a computer. Technically speaking, students could have connected to the Eloqi platform via landlines or cell phones, and could use these devices to speak with trainers; however, it was a long-standing company policy that all participants connect via Eloqi's specially built UI to in order for a lesson to go forward. If this condition wasn't met, the trainers had to terminate the lesson immediately, as in Excerpt 2.1. In it the student (Xia) appears to be unfamiliar with this fundamental condition for participating in a live Eloqi lesson when she reveals that her computer is closed (0:56). The trainer responds by clarifying the expected procedure (1:03) before terminating the lesson, consistent with company protocols. To emphasize, this particular lesson was terminated because the student didn't follow the expected act sequence for accessing an Eloqi trainer, that is, connect to the Eloqi platform via a computer, have the UI open before queuing for the next available trainer, refer to the material on the UI during the lesson with the trainer, etc.

Another crucial procedure for participating in an Eloqi lesson was following the pre-determined lesson plan to the letter. All trainers, no matter their tenure or level of expertise, were required to closely follow the CEL scripts and prompts, as well as the sequence of CEL activities and the allotted time for each. For their part the students were expected to compliantly follow the trainers' cues. From time to time I observed lessons in which students attempted to go off script but, unsurprisingly, trainers generally rebuffed these conversational moves. In Excerpt 2.2 we see just

Excerpt 2.1: You Need to be at a Computer

Amy	Hi. Welcome to LQ English. My name is Amy and I will be your trainer for this session. How are you today, Xia? (..) Hello Xia? (.) Can you hear me? (...) Hello Xia?	0:00
Xia	Hello, hello?	0:27
Amy	Hello, can you hear me?	0:29
Xia	Yes I can.	0:32
Amy	OK, great. Well, welcome to LQ English, and my name is Amy. How are you doing today?	0:33
Xia	Ah, it's fine uh=	0:41
Amy	Good.	0:43
Xia	=(right) now.	0:44
Amy	Good. Well, this morning we are going to do a speaking evaluation and to use LQ English you need a computer. So are you in front of a computer?	0:45
Xia	(.) O::h actually not, no ah, I have just closed my computer.	0:56
Amy	OK ↓ well, you need the com- ah, you need the computer uh, on to do this evaluation, so maybe please give us a call once again when you are at your computer and have it on and ready to go. So, if you have any questions though, you can, ah, call our High Scoring Team and I hope to speak with you, though, sometime. OK?	1:03
Xia	OK ↓	1:26
Amy	Alright, goodbye.	1:27
Xia	Mm goodbye . . . hhh	1:28

such a situation, in which a trainer (Iris) connects with a student (Winson) who requests unstructured conversation.

In the interaction presented in Excerpt 2.2, Winson goes against Eloqi's procedures for participating in a lesson in three ways. First, he reveals that he has not, in fact, connected via the Eloqi UI (0:42) and isn't prepared to follow along on his screen. Second, he has not strategically chosen a CEL lesson to work on, as indicated by his confusion about what lesson he should presumably be doing with Iris right now (1:23–1:56). As paying subscribers, Eloqi students had access to the entire CEL series, the idea being to progress through all the formulas at their own convenience. Students therefore selected which lessons they wanted to do when, and

Excerpt 2.2: I Think We Can Just Talk without the Computer

Iris	Thank you for calling Eloqi English. My name is Iris and I will be your trainer for this session. What's your name?	0:00
Winson	You can- you can call me Winson.	0:09
Iris	OK Winson. How are you doing today?	0:13
Winson	Fine. How are you?	0:16
Iris	I am well. Thank you very much. Um, it looks like we are going to be answering what do you dislike about X type questions today. So let's start by reviewing your pronunciation, alright?	0:18
Winson	OK.	0:32
Iris	OK. You should see a task card on your screen, Winson, I would like you to read the words on it out loud for me, please.	0:33
Winson	A:h but ah, I could not ah see the content on the co— on the screen.	0:42
Iris	OK.	0:48
Winson	Something-	0:49
Iris	=are you having difficulty with your Internet or what's going on?	0:49
Winson	(.) Ah, I think, ah ((clears throat)) I think we can just uh talk, ah, without, ah, the computer-	0:55
Iris	No: I- I'm sorry-	1:02
Winson	(with) the computer (with) the network	1:04
Iris	Yeah, no, I'm sorry, at Eloqi we- we have to work with- with the computer, so you'll need to get your Internet working and then you'll have to call us back.	1:05
Winson	A:h please hold on. Let me try.	1:16
Iris	OK.	1:22
Winson	Ah (..) So could you tell me the name of this lesson?	1:23
Iris	Um, actually you're- you've selected a lesson o:n answering what do you dislike about X type questions. (..) I- I didn't select the lesson- you did. (..) Do you want to get on the Internet and, and go through the lesson first before you talk with us?	1:30
Winson	Ah (..) Let me try again.	1:56
Iris	M'kay. Well, because our interactions are timed, Winson, I'm going to have to let you go until you can get that up and running. So, you do that and then give us a call back. OK?	2:02
Winson	OK uh	2:15
Iris	OK. Thank you.	2:16
Winson	Thanks.	2:19
Iris	Buh-bye.	2:21
Winson	Bye. Bye.	2:22

agreed to do the preparatory activities before connecting with the trainers for live sessions. When Winson admits that he doesn't know what lesson he has selected, he reveals that he has not followed the expected act sequence for engaging in a live lesson with an Eloqi trainer. Finally, in what the trainer reads as the most serious procedural violation, Winson suggests "just talk[ing] without the computer," that is, having a free conversation. The trainer rejects this suggestion, referencing the sanctioned conditions ("we have to work with the computer") as a means of explanation. A few moments later, she takes the decision to end the interaction, again citing expected procedure for doing a lesson properly ("have your computer up and running").

I was working a shift when Iris's lesson with Winson occurred, and I was present in the chat room when she reported this problematic interaction to the supervisor on duty. The other trainers present responded with amusement, as illustrated in Excerpt 2.3.

The surprised and mirthful responses by Iris's supervisor and colleagues revealed the seriousness of this particular procedural breach. Following the lesson scripts was such standard procedure that the trainers could not believe a student would suggest "just chat[ting]." Regardless of Winson's intentions, his actions did not follow the sanctioned procedure for connecting with and participating in an Eloqi lesson, and for these reasons the lesson was terminated and the interaction was marked as failed.

Excerpt 2.3: I Think He Has the Wrong 800# lol

Iris	Disco* with Winson. Said he wasn't on computer and couldn't I	01
	just chat with him. I explained that he needs computer.	02
Supervisor	lol** . . . ok, I informed HST.	03
Daisy	lol @ 'chat with him'	04
Reena	Iris: ROFL*** re: Winson	05
Supervisor	Winson called HST to find out if he could chat with a trainer	06
	without going through a lesson!!!	07
Daisy	Lol	08
Supervisor	they have updated him!!	09
Reena	NUH-UH ROFL	10
Supervisor	Lol	11
Daisy	Does Winson need a friend?	12
Supervisor	lol I think that's a first!!	13
Reena	I think he has the wrong 800#**** lol	14
Daisy	Lol	15

* Disconnect

** Laughing out loud

*** Rolling on the floor laughing

****A reference to toll-free phone numbers starting with the digits 1-800.

NAVIGATION: HOW TO NAVIGATE THE ELOQI UI

As I myself discovered when I worked as an Eloqi trainer, competently participating in the lessons required close attention to numerous details presented on the UI, many of them time-sensitive. Eloqi's proprietary UI was constantly being tweaked, refined, and updated in response to trainer feedback and in support of the company's long-term technical and business plans. The technical team regularly introduced new tools and features while the manager of the trainer team and the content developer instructed the trainers in the corresponding policies, guidelines, and tips for their use. The trainers used the in-house forum to actively discuss the effective use of the UI, covering popular topics like how to use hot keys to type up feedback faster.

Perhaps unsurprisingly, given the complexity of the UI, one class of problematic trainer-student communication pertained to procedures for using the Eloqi UI effectively during lessons. In Excerpt 2.4 for example, we see a trainer (Carly) struggling to teach a student (Jacqueline) how to utilize the chat window feature.

Here the trainer attempts to teach the student a new vocabulary word, "specific," by typing it into the chat window where the student will be able to see it. The trainer's repeated efforts to direct the student's attention to the chat window (11:59, 12:08, 12:21, 12:33, 12:41) combined with the student's perplexed responses and silences indicate the student's momentary confusion about what the chat window is and how it should be used in this context. A full two minutes elapse until the trainer and student establish that they are both looking at the same thing on the UI (13:47) and by this time the interaction is nearing the maximum time of fifteen minutes. The trainer briefly explains the procedure for using the chat window (13:56, 14:12) but shortly thereafter begins to recite the closing statements before ending the call, thereby staying within the time limit for the lesson.

Excerpt 2.4 illustrates how Eloqi trainer-student interactions could stall when either participant—but most commonly the student—was unfamiliar with the features of the UI and/or the procedures for using them during a live interaction. Regardless of the underlying reason for the confusion (terminology, being a novice user, language barriers, etc.), not knowing the procedure for using a UI feature could slow down or even bring the lesson to a halt. Furthermore, because of the strict time limit for these lessons (fifteen minutes), slowed or halted progression through the lesson was a serious problem for both parties.

Excerpt 2.4: Do You See the Chat Window?

Carly	U:m, do you know the word "specific"? I'm going to put it in the chat window. Specific.	11:46
Jacqueline	Spe-ci-city hhh . . .	11:55
Carly	Do- yeah, so, Ja-	11:57
Jacqueline	(sorry)	11:58
Carly	Jacqueline, do you see the chat window on the left hand side?	11:59
Jacqueline	Hat window?	12:06
Carly	Yeah. Do you see the chat window on the left of your screen?	12:08
Jacqueline	(.) Sorry I hhh . . .	12:18
Carly	That- that's OK, that's OK. On the left side of the screen there is a chat window (.) and I'm ty-	12:21
Jacqueline	Uh, chat window.	12:32
Carly	Yeah, and I'm <u>typing</u> some words there.	12:33
Jacqueline	(.) Oh.	12:39
Carly	Uh, can you see the words?	12:41
Jacqueline	Uh, no.	12:44
Carly	You can't. Are you sitting by the computer?	12:47
Jacqueline	Yeah, I'm sitting in front of computer.	12:52
Carly	OK. And then do you see the- the screen? (..) Can you see the- the interaction screen?	12:57
Jacqueline	Inter ° (action scr)° (.) Ah=	13:09
Carly	OK, OK	13:14
Jacqueline	= Oh- oh- Oh. Sorry hhh . . .	13:15
Carly	OK. That's OK. Don't worry. Um, so when you use Eloqi, ah, we can talk to each other and we can send each other messages. So right now I am sending you a message. I'm typing a message. Can you see the message?	13:18
Jacqueline	(.) Uh, OK, I- Oh. I see that.	13:47
Carly	You <u>see</u> it?	13:54
Jacqueline	Yes.	13:55
Carly	OK, good. OK. So sometimes if there is a word that, that, um, I want to teach you, I can put it in this text message.	13:56
Jacqueline	(..) Oh.	14:11
Carly	Ah, so I put some vocabulary there for you.	14:12
Jacqueline	(.) Oh yeah.	14:19

TASK/ACTIVITY CONTENT:
HOW TO PROCEED THROUGH AN ELOQI LESSON

Among the examples of problematic communication during trainer-student interactions, the most common type was that in which students misunderstood the act sequence for completing specific speaking tasks and activities. As previously mentioned, Eloqi had fixed lesson plans, not to be deviated from, and there was a pre-sequenced set of activities to complete during each fifteen-minute interaction. I found numerous cases of students not understanding the company's pre-determined procedure for the particular tasks at hand. For example, in Excerpt 2.5 the trainer (Daisy) and the student (Grace) are practicing the formula for answering the question type "How often do you do ~?" They have completed the pronunciation practice and now begin a section in which the student must utilize material listed on the task card (a visual prompt) to respond to the trainer's questions. The task card lists sample activities (eat Western food, swim in the sea, read books) and the following adverbs of frequency: *rarely, occasionally, frequently, every day, once in a blue moon, never,* and *almost never.*

In Excerpt 2.5, the trainer introduces the activity by way of reading the provided script (4:09), thereby calling attention to the standard Eloqi act sequence for this task:

1. The trainer (Daisy) will show the student the visual cue (task card), which lists activities and adverbs of frequency.
2. The trainer will pose questions to the student. Though the trainer doesn't explicitly say so in advance of the activity, all of the questions will be about the activities listed on the card.
3. After listening to each question, the student must provide an answer using one of the adverbs of frequency listed on the card. The student's answers should be one to two sentences long, and they should be accurate. (Later in the interaction the trainer adds that the answers must also be full sentences.)

Although the student's first answer does not incorporate any of the listed adverbs of frequency (5:06), the trainer does not correct her orally but rather proceeds on to the next question (5:18). Again the student answers with an adverb of frequency (*once a month*) that is not listed on the task card. After a long pause, the trainer reemphasizes the procedure and adds another stipulation: answers must be given in full sentences (5:49). What follows is a drawn out exchange during which the trainer repeatedly attempts to explain the procedure, giving explicit directives in six separate conversational turns. More than five minutes elapse before the student

Excerpt 2.5: Answer the Question Using the Adverbs of Frequency

Daisy	OK now let's practice the language you'll need to answer the IELTS type questions for this lesson.	4:02
Grace	(.) °OK°	4:08
Daisy	OK first let's look at the adverb of frequency. I will show you a task card with different activity- activities and adverbs of frequency. Please listen to my questions, and answer the questions with one or two short accurate sentences. OK?	4:09
Grace	OK. ((clears throat loudly)) (..)	4:26
Daisy	Do you see the task card?	4:35
Grace	(.) Ah yeah. I see. (..)	4:37
Daisy	OK, how often do you go out to sing kar()? (. . .)	4:46
Grace	Ah pardon? (..)	4:54
Daisy	How often do you go out to sing karaoke?	5:01
Grace	Um. Ah. I often, um, go out to sing karaoke, ah, (every weeks).	5:06
Daisy	(..) And how often do you eat Western food?	5:18
Grace	((clears throat loudly)) mm uh usually mm I uh (let me see) uh, once a mo↑nth (. . .)	5:25
Daisy	OK. Can you answer the questions using the information on the task card, please, in a full sentence?	5:49
Grace	Ah yeah, I see. (. . .)	5:56
Daisy	Gra:ce?	6:14
Grace	Ah yeah.	6:16
Daisy	How <u>often</u> do you <u>eat</u> Western <u>food</u>?	6:19
Grace	Um:: Ah, to be honest I don't like, ah, eat Western food. Ah, ma:ybe several, ah, several months, ah, I, I, I go out, to, ah, eat Western food.	6:23
Daisy	(..) OK. So can you-	6:47
Grace	(Hello?) Oh. OK.	6:48
Daisy	How would you answer the question- how would you answer the question using the adverbs of frequency and the activities on your task card?	6:52
Grace	Um. (.) (. . .)	7:01
Daisy	Grace?	7:36
Grace	Ah, yeah. I'm here. (.) Hello?	7:38
Daisy	Do you- do you see the adverb of frequency?	7:44
Grace	(..) Of frequency.	7:48
Daisy	Are you looking at your task card?	7:52
Grace	Ah, yeah	7:55

(continued)

Excerpt 2.5. *(continued)*

Daisy	OK. I need you to answer how often do you eat Western food using the adverbs of frequency and activity on your task card, please.	7:57
Grace	Ah, so- can you- can you- u:m (.) I have- I have answer the question.	8:14
Daisy	That's not correct. I need you to use the information on the task card to properly answer the question.	8:24
Grace	O:h (..) I must use the words, um, left to right. (. . .)	8:34
Daisy	OK. I need you to use a full sentence and use the adverb of frequency and the activity on your student ca:rd to answer the question how often do you eat Western food.	8:47
Grace	Um. Hhh . . . °frequency° I- I eat Western food frequency.	9:04
Daisy	OK Gra:ce, do you see the adverb of frequency list? Rarely, occasionally, frequently, everyda:y	9:17
Grace	every day	9:28
Daisy	Once in a blue moon, never, almost never. Do you see that list?	9:29
Grace	Ah. Ye:ah. I see.	9:35
Daisy	OK. I need you to use that list to answer the questions that I am asking you. So using a word from that list, tell me how often you eat Western food?	9:37
Grace	(.) Uh frequency. (. . .)	9:53
Daisy	OK. Do you eat Western food rarely, occasionally, frequently, every day, once in a blue moon, never, almost never. How often do you eat Western food?	10:08
Grace	Um: I eat Western food, ah, frequen(cy).	10:23
Daisy	Frequently.	10:29
Grace	Frequently.	10:30
Daisy	Frequently.	10:32
Grace	Frequently.	10:36
Daisy	OK. Now how often do you go swimming in the sea?	10:37
Grace	U:h (..) rarely.	10:45
Daisy	OK, and full sentence, please.	10:53
Grace	(.) I beg your pardon?	10:58
Daisy	I need you to answer these questions in a full sentence, please.	11:00
Grace	U:m I, I go swimming in the sea rarely uh because I- I have not enough time to go- uh to the sea.	11:07
Daisy	OK. So, I rarely go swimming in the sea.	11:24
Grace	(uh) rarely go swimming in the sea.	11:29
Daisy	Now how often do you do physical exercise?	11:33
Grace	Mm: ah I do physical exercise every day, ah, when I finish my ah cla- uh class (mostly) I- I always (run) to, mm playground and do some, mm, sports, ah, like jogging, um, mm..	11:39
Daisy	OK. So your answer, Grace, would simply be, I do physical exercise every day after class. OK?	12:06

Grace	OK.	12:16
Daisy	Alright. So how often do you read novels?	12:17
Grace	(.) Uh, to be honest, ah, ah, almost never, um, because I think that, ah, reading is boring.	12:24
Daisy	(.) OK. So I almost-	12:42
Grace	(.)	
Daisy	I almost never read novels because I think reading is boring.	12:44
Grace	Yeah.	12:49
Daisy	(.) OK. Do you understand what I did with those?	12:53
Grace	Yes. I un (.)	12:58
Daisy	OK. Alright. Now I am going to show you another task card and ask you what you usually do at different times of the day. So we can work on the present tense and do a little bit more adverbs of frequency. OK?	13:00

produces the desired type of answer at 11:07. Considering that five minutes is a full one-third of the allotted time for the lesson, this lengthy exchange in clarifying the activity procedure has cost significant resources.

TROUBLESHOOTING: HOW TO HANDLE TECHNICAL ISSUES DURING AN ELOQI LESSON

The final category of procedural issues in the data set pertained to handling technical issues that arose during the one-to-one sessions between trainers and students. The most common type of technical issue at Eloqi was sound problems. It was not uncommon to experience degradation in the audio (words sounding blurred or slurred, choppiness, sound dropping out altogether, etc.) caused by weaknesses in the Internet connection. Other sound problems like echoing (often caused by one or both speakers not wearing a headset), pronounced volume variation, and static were also par for the course. When sound issues became so troublesome that they caused significant disruption to the lesson, the trainers were permitted to terminate the call, ideally after directing the student to call HST for assistance. Finally, the trainers would report the technical issue to the supervisor on duty in the chat room.

In theory, the procedure for handling technical difficulties was straightforward, but in practice it often became muddled, as in Excerpt 2.6. In it, the trainer (Iris) is halfway through the lesson with the student (Lei) when she notices an echo on the line. Iris identifies the problem and attempts to troubleshoot it with the student. She calls the student's attention to the issue and issues a vague directive (7:43) followed by a clearer one (7:58). Over the following turns the trainer makes repeated references to the problem but the student appears not to understand either the trainer's identification of the problem or her instructions about dealing with it. At 8:52 the trainer advances to the standard procedure for such cases, telling the student that they must end the call, and that the student should check in with HST. While we can't be sure if the student understands that the trainer is complaining about an echo, she does appear to be familiar with the standard procedure for disconnecting and calling HST, and indicates agreement to take these actions (9:10). However, at 9:25 the trainer finds that the echo has receded and changes the plan, offering to continue the lesson. Understandably, the student is puzzled about what should happen next (9:45 and 10:01) despite the trainer's prompting (9:43, 9:57). It takes several more turns for the trainer and student to arrive at a mutual understanding about carrying on with the lesson.

In this case, the act sequence for identifying a technical issue is arduous and unsuccessful, as there is no clear indication that the student has understood either the problem (echo) or the procedure for dealing with it (re/plug in the headset). The trainer's attempts to have the student resolve the technical issue prove to be fruitless as the steps followed by the trainer are—at least initially—unfamiliar to the student. It is only when the trainer falls back on the standard procedure for troubleshooting (end the interaction, call HST) that mutual understanding is reached, but this mutual understanding is upset when the trainer veers away from the agreed-upon procedure.

DISCUSSION

Despite Eloqi's attempts to systematize and control trainer-student communication by implementing a detailed lesson protocol, there were—perhaps inevitably—cases of problematic and sometimes failed communication. In analyzing these cases, I found the act sequence variable from Hymes's SPEAKING heuristic (Hymes 1964, 1972a) combined with the related construct of procedural knowledge to be very useful for understanding how and why this problematic communication between Eloqi trainers and students occurred. Through an EC-based analysis of the cases I was able to sort the problematic communication into the following four categories:

1. Initiation and participation procedures—how to take part in an Eloqi lesson
2. Navigation procedures—how to navigate the UI
3. Task procedures—how to proceed through a task or activity
4. Troubleshooting procedures—how to handle technical issues

Excerpt 2.6: Can You Get Rid of that Echo, Please?

Iris	Ok, so let's look at future ambition phrases, and here is the 3 steps. *((Her voice echoes in the background.))*	07:35
Lei	Mmhm.	07:42
Iris	Um, I ca- uh, right now Lei, I am hearing an echo of my voice. Can you get rid of that echo, please?	07:43
Lei	Uh, s- sorry, could you uh- could you speaking? One time?	07:53
Iris	Lei, I am hearing an echo of my voice and I can't hear you clearly. Are you using um, a headset, and if you are, could you plug it in, please?	07:58
Lei	My phone is not- is unclear?	08:11
Iris	There's an <u>echo</u>=	08:15
Lei	Echo.	08:17
Iris	=I hear my voice, and your voice.	08:18
Lei	O:::h. No, I listen clearly.	08:22
Iris	Ok well that's great, but I am not able to listen clearly.	08:27
Lei	Ok.	08:33
Iris	Are you using your computer or are you using a telephone?	08:34
Lei	No, I don't- I don't use the telephone.	08:38
Iris	Ok, so I need you to plug in your headset, so I don't hear the echo.	08:42
Lei	Oh- OK.	08:51
Iris	Ok. *((voice continues to echo))* Ok, I am still hearing that echo. Lei, I am going to ask that you call our high scoring team and have them troubleshoot an echo sound with you. Ok?	08:52
Lei	Ok.	09:04
Iris	Call them and tell them 'my trainer said that there is an echo, can you help me?' *((echoing sound seems to recede))*	09:05
Lei	Oh, uh ye- (now) I can hear you. I:- I will- mm I can ( ) the ( ) on the (Skype) with the LQ English high (scoring) team.	09:10
Iris	Alright, I- I don't know what you just said but the echo has gone away so let's take a look at the future ambition phrases on your screen. If the echo comes back, I am going to hang up the call and you're going to call HST for help, OK?	09:25
Lei	Ok.	09:42
Iris	Ok. Can you see the card on your screen?	09:43
Lei	Uh, just a moment. (.) Yeah, I can see.	09:45
Iris	Ok:: go ahead and begi↑n. .	09:57
Lei	Ok. (..) (I will) call the high (circum) team phone number.	10:01
Iris	(.) Um, if you want to call high scoring team, I am going to have to disconnect our ca↓ll or you can try the card that's in front of you=	10:18
Lei	OK	
Iris	=Did you wanna go ahead and do the exercise?	10:26
Lei	Yeah, I:: I hope- I hope to continue to (stay) uh continue to talking with you.	10:28
Iris	Ok well then go ahead and do the exercise that's on your computer screen.	10:39

Here I will discuss the larger implications of these findings, focusing on their relation to UX and interaction design.

UIs are a means not only of presenting information, options, and activities to the users, but also of organizing information, options, and activities. In this way they are implicated in users' interpretational, sense-making, and decision-making processes (Beer 2008; Gane and Beer 2008; Manovich 2001, 2003). In Eloqi's case, the design of the UI lays out a very deliberate procedure for trainer-student communication, and it directly guides users through the lessons in the manner determined by the organization to be valid. The UI prompts trainers and students on what speech acts (greeting, asking, telling, saying, giving information, correcting, checking, clarifying, challenging, clicking, directing, saying goodbye, etc.) to perform in what sequence, and for what length of time. These prompts simultaneously demonstrate what counts as legitimate communication for these speakers (Eloqi trainers and students) in this context (live Eloqi lesson). Through the force of the community's agreed-upon rules (follow the scripts, stay within the time limits) the UI curtails the options for speech. In these ways, the UI actually encodes Eloqi's expectations for competent communicative behaviors during a live English lesson.

Encoding Eloqi's UI with cues for competent communication was not accidental. On the contrary, it was precisely the intention of Eloqi's engineers who, in concert with the company's visionaries, designed an approach to online communication training that they felt was scalable and amenable to mass reproduction without significant variation or loss of quality. The success of this design rested in large part on shared understandings of *procedure*, that is, a set of explicit, sequenced communicative acts which, when performed according to local expectations, comprised competent behavior during a live Eloqi lesson. Eloqi was able to make some of its locally required procedures visible in the UI, but for other procedures it took time, training, and practice for them to become intuitive. In other words, these procedures were not sufficiently encoded to allow for maximum agency, as defined by Murray (2012) earlier in this chapter.

People develop procedural knowledge over time, through socialization, experience, and repetition. We enter into communication situations, technology-mediated or otherwise, with cognitive scripts already in mind (Shoemaker 1996). Simultaneously, we test and adjust those scripts in our moment-to-moment interactions, storing our developing procedural knowledge for future reference and use. As we experience new situations we recall this knowledge and use it accordingly as we interpret and respond to communicative situations (Gioia and Poole 1984). Over time novices learn locally expected procedures and can intuitively or automatically engage in the communication at hand (Cameron 2000a, 2008, 2000b).

Learning the procedure for a communicative activity is thus a work in progress.

This process of learning the procedure for a technology-mediated communicative activity must be of special interest to UX designers, who can benefit from exploring how users draw on extant procedural knowledge pertaining to routine tasks to make sense of new technology-mediated spaces in which they are engaging in novel activities (Sternberg 2009; cf. Boellstorff 2008; Kendall 2002). Indeed, as I learned while conducting this research at Eloqi, all of the trainers and students had experience in teaching and/or learning, all had spent some part of their lives participating in their country's formal education system. They must have used their knowledge of engaging in traditional (offline) learning settings as they navigated Eloqi's virtual learning community and engaged in the company's unique teaching and learning activities. What gave this process special urgency in the Eloqi community were the constraints that the company built into the interaction design, particularly the strict time limits placed on the trainers and students throughout their interactions. Because of this, sustained misunderstandings about the expected procedures were costly to Eloqi's members and potentially wasted a limited resource: time. For these reasons, it was critical that Eloqi's users pick up the locally expected procedures as quickly as possible.

Taken as a whole, it makes sense in all phases of the design process to highlight the concepts of act sequence and procedural knowledge; doing so draws our attention to the "what happens now and what happens next" components of technology-mediated interactions from both the design and use perspectives. The procedures and act sequences designed for a UI must adequately fit the needs and goals of the organizations commissioning the UI, the boots-on-the-ground service providers or representatives, the clients, and the affordances and constraints of the technological platform itself. Technological interfaces are "culturally defined, which means that generally, the social meaning of an interface is not always developed when the technology is first created but usually comes later, when it is finally embedded in social practices" (de Souza e Silva 2006, 261–262). Because of this, it is beneficial to examine local notions of act sequence and procedural knowledge not only at the start of the design process, but throughout the life cycle (design, creation, launch, use, redesign, ongoing use) of the build.

NOTES

1. Publisher's Note: The screen shots, references, and information pertaining to the company identified by the pseudonym Eloqi is published with the company's

knowledge and agreement that the screen shots, references, and information would be used in a later publication. Likewise, the interviews used as supplemental research in this text were all conducted with the participants' knowledge and agreement that these interviews would be used in a later publication.

2. Pseudonyms have been applied to the company and all of its members (admins, trainers, students) in order to protect their privacy.

3. Consider how expert we can be at using the grammar of our native language while not being able to explain it to a non-native speaker.

4. Going over the fifteen-minute limit was cause for reprimand, and if a trainer repeatedly failed to stay within the time constraints, they could be dismissed.

REFERENCES

Aakhus, Mark, and Sally Jackson. "Technology, Design, and Interaction." In *Handbook of Language and Social Interaction*, edited by Kristine L. Fitch and Robert E. Sanders, 411–36. Mahwah, NJ: Erlbaum, 2005.

Appel, Christine, Jackie Robbins, Joaquim More, and Tony Mullen. "Task and Tool Interface Design for L2 Speaking Interaction Online." Paper presented at the EUROCALL Conference, Gothenburg, Sweden, August 22–25, 2012.

Bailey, Benjamin. "Communication of Respect in Interethnic Service Encounters." *Language in Society* 26, no. 3 (1997): 327–56.

Baxter, Leslie. "'Talking Things Through' and 'Putting It in Writing': Two Codes of Communication in an Academic Institution." *Journal of Applied Communication Research* 21 (1993): 313–26. doi:10.1080/00909889309365376.

Beer, David. "The Iconic Interface and the Veneer of Simplicity: Mp3 Players and the Reconfiguration of Music Collecting and Reproduction Practices in the Digital Age." *Information, Communication & Society* 11, no. 1 (2008): 71–88.

Boellstorff, Tom. *Coming of Age in Second Life: An Anthropologist Explores the Virtually Human*. Princeton, NJ: University Press, 2008.

Boromisza-Habashi, David, and Russell M. Parks. "The Communal Function of Social Interaction on an Online Academic Newsgroup." *Western Journal of Communication* 78, no. 2 (2014): 194–212.

Cameron, Deborah. *Good to Talk? Living and Working in a Communication Culture*. London: Sage, 2000a.

Cameron, Deborah. "Styling the Worker: Gender and the Commodification of Language in the Globalized Service Economy." *Journal of Sociolinguistics* 4, no. 3 (2000b): 323–47.

Cameron, Deborah. "Talk from the Top Down." *Language & Communication* 28 (2008): 143–55.

Carbaugh, Donal. *Talking American: Cultural Discourses on Donahue*. Norwood, NJ: Ablex, 1988a.

Carbaugh, Donal. *Cultures in Conversation*. Mahwah, NJ: Lawrence Erlbaum Associates, Inc., 2005.

Carbaugh, Donal. "Cultural Discourse Analysis: Communication Practices and Intercultural Encounters." *Journal of Intercultural Communication Research* 36, no. 3 (2007): 167–182. doi:10.1080/17475750701737090.

Carbaugh, Donal, Ute Winter, Brion van Over, Elizabeth Molina-Markham, and Sunny Lie. "Cultural Analyses of in-Car Communication." *Journal of Applied Communication Research* 41, no. 2 (2013): 195–201.

Cooper, Alan. *The Inmates Are Running the Asylum: Why High Tech Products Drive Us Crazy and How to Restore the Sanity.* Indianapolis, IN: Sams Publishing, 2004.

Coutu, Lisa. "Communication Codes of Rationality and Spirituality in the Discourse of and about Robert S. Mcnamara's 'in Retrospect.'" *Research on Language and Social Interaction* 33, no. 2 (2000): 179–211.

Coutu, Lisa. "Contested Social Identity and Communication in Talk and Text About the Vietnam War." *Research on Language & Social Interaction* 41, no. 4 (2008): 387–407.

de Souza e Silva, Adriana. "From Cyber to Hybrid: Mobile Technologies as Interfaces of Hybrid Spaces." *Space and Culture* 9, no. 3 (2006): 261–78.

Dix, Alan, Janet E. Finlay, Gregory D. Abowd, and Russell Beale. *Human-Computer Interaction*, 3rd edition. Essex, England: Pearson Education Limited, 2004.

Dori-Hacohen, Gonen, and Nimrod Shavit. "The Cultural Meanings of Israeli Tokbek (Talk-Back Online Commenting) and Their Relevance to the Online Democratic Public Sphere." *International Journal of Electronic Governance* 6, no. 4 (2013): 361–79.

Edgerly, Louisa. "Difference and Political Legitimacy: Speakers' Construction of 'Citizen' and 'Refugee' Personae in Talk About Hurricane Katrina." *Western Journal of Communication* 75, no. 3 (2011): 304–22.

Emerson, Robert M., Rachel I. Fretz, and Linda L. Shaw. *Writing Ethnographic Fieldnotes*. Chicago, IL: The University of Chicago Press, 1995.

Fong, Mary. "'Luck Talk' in Celebrating the Chinese New Year." *Journal of Pragmatics* 32 (2000): 219–37.

Gane, Nicholas, and David Beer. *New Media: The Key Concepts*. Oxford: Berg, 2008.

Gioia, Dennis A., and Peter P. Poole. "Scripts in Organizational Behavior." *Academy of Management Review* 9, no. 3 (1984): 449–59.

Hart, Tabitha. "Speech Codes Theory as a Framework for Analyzing Communication in Online Educational Settings." In *Computer Mediated Communication: Issues and Approaches in Education*, edited by Sigrid Kelsey and Kirk St. Amant. Hershey, PA: IGI Global, 2011.

Hart, Tabitha. "Learning How to Speak Like a 'Native': A Case Study of a Technology-Mediated Oral Communication Training Program." *Journal of Business and Technical Communication* (2016).

Holdford, David. "Service Scripts: A Tool for Teaching Pharmacy Students How to Handle Common Practice Situations." *American Journal of Pharmaceutical Education* 70, no. 1 (2006): 1–7.

Huspek, Michael. "Oppositional Codes and Social Class Relations." *The British Journal of Sociology* 45, no. 1 (1994): 79–102.

Hymes, Dell. "The Ethnography of Speaking." In *Anthropology and Human Behavior*, edited by Thomas Gladwin and William Sturtevant, 13–53. Washington, DC: The Anthropological Society of Washington, 1962.

Hymes, Dell. "Toward Ethnographies of Communication." *American Anthropologist* 66, no. 6 (1964): 1–34.

Hymes, Dell. "Models for the Interaction of Language and Social Life." In *Directions in Sociolinguistics: The Ethnography of Communication*, edited by John Gumperz and Dell Hymes, 35–71. New York: Basil Blackwell Inc., 1972a.

Hymes, Dell. "On Communicative Competence." In *Sociolinguistics*, edited by J. B. Pride and Janet Holmes, 269–85. Baltimore, MD: Penguin Education, 1972b.

Hymes, Dell. *Foundations in Sociolinguistics: An Ethnographic Approach*. Philadelphia, PA: University of Pennsylvania Press, 1974.

Jackson, Sally, and Mark Aakhus. "Becoming More Reflective About the Role of Design in Communication." *Journal of Applied Communication Research* (2014): 1–10.

Katriel, Tamar. *Talking Straight: Dugri Speech in Israeli Sabra Culture*. Cambridge, UK: Cambridge University Press, 1986.

Katriel, Tamar, and Gerry Philipsen. "'What We Need Is Communication': 'Communication' as a Cultural Category in Some American Speech." *Communications Monographs* 48 (1981): 301–17.

Keating, Elizabeth. "The Ethnography of Communication." In *Handbook of Ethnography*, edited by Paul Atkinson, Amanda Coffey, Sara Delamont, John Lofland, and Lyn Lofland, 285–300. London: Sage, 2001.

Kendall, Lori. *Hanging out in the Virtual Pub: Masculinities and Relationships Online*. Berkeley: University of California Press, 2002.

Leighter, James L., and Laura Black. "'I'm Just Raising the Question': Terms for Talk and Practical Metadiscursive Argument in Public Meetings." *Western Journal of Communication* 74, no. 5 (2010): 547–69.

Leighter, James L., Lisa Rudnick, and Theresa J. Edmonds. "How the Ethnography of Communication Provides Resources for Design." *Journal of Applied Communication Research* 41, no. 2 (2013): 209–15.

Manovich, Lev. *The Language of New Media*. Cambridge, MA: MIT Press, 2001.

Manovich, Lev. "New Media from Borges to Html." In *The New Media Reader*, edited by Noah Wardrip-Fruin and Nick Montfort. Cambridge, MA: MIT Press, 2003.

McKay, Everett N. *UI is Communication: How to Design Intuitive, User Centered Interfaces by Focusing on Effective Communication*. Waltham, MA: Elsevier, 2013.

Milburn, Trudy. "S.P.E.A.K.I.N.G.: A Research Tool." *Communication Institute for Online Scholarship (CIOS)*. Accessed January 1, 2015. http://www.cios.org/encyclopedia/ethnography/4speaking.htm.

Murray, Janet H. *Inventing the Medium: Principles of Interaction Design as a Cultural Practice*. Cambridge, MA: The MIT Press, 2012.

Nass, Clifford, Jonathan Steuer, and Ellen R. Tauber. "Computers Are Social Actors." Paper presented at the CHI, 1994.

Nickols, Fred. "The Knowledge in Knowledge Management." In *The Knowledge Management Yearbook 2000–2001*, edited by John A. Woods and James Cortada, 12–21. Boston, MA: Butterworth-Heineman, 2000.

Nielsen, Jakob. "Heuristic Evaluation." In *Usability Inspection Methods*, edited by Jakob Nielsen and Robert L. Mack, 25–62. New York: John Wiley & Sons, Inc., 1994.

Nielsen, Jakob. "10 Usability Heuristics for User Interface Design." Accessed January 2, 2015. http://www.nngroup.com/articles/ten-usability-heuristics/.

Philipsen, Gerry. "Speaking 'Like a Man' in Teamsterville: Culture Patterns of Role Enactment in an Urban Neighborhood." *Quarterly Journal of Speech* 61, no. 1 (1975): 13–23.

Philipsen, Gerry. *Speaking Culturally: Explorations in Social Communication*. Albany, NY: State University of New York Press, 1992.

Philipsen, Gerry. "A Theory of Speech Codes." In *Developing Communication Theories*, edited by Gerry Philipsen and Terrance L. Albrecht, 119–56. New York: State University of New York Press, 1997.

Philipsen, Gerry. "Permission to Speak the Discourse of Difference: A Case Study." *Research on Language & Social Interaction* 33, no. 2 (2000): 213–34.

Philipsen, Gerry. "Cultural Communication." In *Handbook of International and Intercultural Communication*, edited by William B. Gudykunst and Bella Mody, 51–67. Thousand Oaks, CA: Sage, 2002.

Philipsen, Gerry. "Some Thoughts on How to Approach Finding One's Feet in Unfamiliar Cultural Terrain." *Communication Monographs* 77, no. 2 (2010): 160–68.

Philipsen, Gerry, and Lisa M. Coutu. "The Ethnography of Speaking." In *Handbook of Language and Social Interaction*, edited by Kristine L. Fitch and Robert E. Sanders, 355–79. Mahwah, NJ: Erlbaum, 2005.

Philipsen, Gerry, Lisa M. Coutu, and Patricia Covarrubias. "Speech Codes Theory: Restatement, Revisions, and Response to Criticisms." In *Theorizing About Intercultural Communication*, edited by William Gudykunst, 55–68. Thousand Oaks, CA: Sage, 2005.

Philipsen, Gerry, and James L. Leighter. "Sam Steinberg's Use of 'Tell' in After Mr. Sam." In *Interacting and Organizing: Analyses of a Management Meeting*, edited by Francois Cooren, 205–23: Lawrence Erlbaum Associates, Inc., 2007.

Saville-Troike, Muriel. *The Ethnography of Communication: An Introduction*. Baltimore, MD: University Park Press, 1982.

Saville-Troike, Muriel. *The Ethnography of Communication: An Introduction*, 3rd edition. Malden, MA: Blackwell Publishing, 2003.

Shoemaker, Stowe. "Scripts: Precursor of Consumer Expectations." *Cornell Hotel and Restaurant Administration Quarterly* (February 1996): 42–53.

Sprain, Leah, and David Boromisza-Habashi. "The Ethnographer of Communication at the Table: Building Cultural Competence, Designing Strategic Action." *Journal of Applied Communication Research* 41, no. 2 (2013): 181–187.

Sprain, Leah, and John Gastil. "What Does It Mean to Deliberate? An Interpretive Account of Jurors' Expressed Deliberative Rules and Premises." *Communication Quarterly* 61, no. 2 (2013): 151–71.

Sternberg, Janet. "Misbehavior in Mediated Places: Situational Proprieties and Communication Environments." *ETC: A Review of General Semantics* 66, no. 4 (2009): 433–22.

Vorvoreanu, Mihaela. "Perceptions of Corporations on Facebook: An Analysis of Facebook Social Norms." *Journal of New Communications Research* 4, no. 1 (2009): 67–86.

Winchatz, Michaela R. "Social Meanings in German Interactions: An Ethnographic Analysis of the Second-Person Pronoun Sie." *Research on Language and Social Interaction* 34, no. 3 (2001): 337–69.

Witteborn, Saskia. "Communicative Competence Revisited: An Emic Approach to Studying Intercultural Communicative Competence." *Journal of Intercultural Communication Research* 32, no. 3 (2003): 187–203.

Witteborn, Saskia. "Discursive Grouping in a Virtual Forum: Dialogue, Difference, and the 'Intercultural.'" *Journal of International and Intercultural Communication* 4, no. 2 (2011): 109–26. doi:10.1080/17513057.2011.556827.

Witteborn, Saskia. "Forced Migrants, New Media Practices, and the Creation of Locality." In *The Handbook of Global Media Research*, edited by Ingrid Volkmer, 312–30. Malden, MA: Blackwell Publishing Ltd., 2012.

Witteborn, Saskia, Trudy Milburn, and Evelyn Y. Ho. "The Ethnography of Communication as Applied Methodology: Insights from Three Case Studies." *Journal of Applied Communication Research* 41, no. 2 (2013): 188–94.

Witteborn, Saskia, and Leah Sprain. "Grouping Processes in a Public Meeting from an Ethnography of Communication and Cultural Discourse Analysis Perspective." *The International Journal of Public Participation* 3, no. 2 (2009): 14–35.

Wolcott, Harry F. *Ethnography: A Way of Seeing*. Walnut Creek, CA: AltaMira Press, 1999.

II

INTERACTING AND RELATING
Trudy Milburn

Users' interactions with and through digital media are not trouble-free. The two studies in the previous section provided examples illustrating the way communicative acts, such as directives or giving instruction can lead to misunderstandings in certain situations. The designation of who can say what to whom is an important consideration for both conversation participants as well as researchers who use this information to make interpretations based on specific acts or when considering entire episodes of interaction. By closely examining interactional episodes, researchers make interpretations about the types of relationships people are creating with digital media as well as with one another. For one, if people act towards machines as if they were people (Murray 2012), we need to know much more about how typical conversations for routine matters are undertaken. Therefore, in order to help developers and designers learn more about how the tasks accomplished through digital media influence relationships, researchers in this section address the question, "how does the way people interact with a specific digital medium in a particular situation impact their relationships?"

When considering the way interaction forms relationships, it is useful to refer back to interpersonal communication findings. Sigman (1995) explains that relationships are both a "category of meaning" (190) and a "communicative achievement" (191) based on the patterned practices of members of a community. He further suggests examining the way relational practices "identif[y] people," their associations, as well as their "rights and responsibilities . . . across space and time" (190). This perspective of relationships locates them in the ongoing stream of interaction. Sigman differentiates between two types of relationships—those that he calls social relationships—that transcend interactional moments, and what he calls interaction-order relationships, those that are created and maintained within particular interactional episodes, like questioner-answerer (193). While not confined to particular interactions, relationships generate a set of obligations. Sigman advocates using a three-part analytic framework for examining relationships:

> (1) the sociocultural repertoire permitting and constraining behavior and defining each relationship category; (2) the communication activities that accomplish, negotiate, and orient to these resources; and (3) the character, composition, and continuity of the relationships across selected episodes and time frames that are built on these communication activities. (198)

Extending the consideration of interpersonal relationships at the dyadic level from those that operate with more complexity at the group level, Keyton (1999) has explored the ways that relationships between group members may be distinct from the relationship that develops within the whole group itself. Group relationships develop where norms and shared meaning occur in repeated interactions over time. Keyton also warns us not to confuse task-oriented group research from groups that form for more social purposes, and to consider the way tasks and social relationships are mutually influential. Put another way, social and task dimensions are present in all groups, regardless of their primary purpose for coming together (Ellis and Fisher 1994).

Up to now, we have alluded to designers who refers to "interaction" as part of a problem to be solved, including, "the many aspects of the system that have to be the subject of coordinated design decisions, including social and cultural elements as well as technical and visual components" (Murray 2012, 11). The term "interaction design" suggests a purposeful construction of interaction options made available to eventual users. While the previous section considered the ways that interaction connotes both a single individual using a digital medium, as well as interaction with others through the use of digital media, this section is focused exclusively on the latter.

LSR studies operate under the assumption that interaction is a communicative practice that is both structurally constituted by and normatively reified within each enacted episode. In addition, through the practice of communication within interactions, people are simultaneously forming and/or maintaining relationships. The chapters in this next section describe several types of relationships built and maintained through and/or in the use of digital media, including co-workers, teacher-student relationships, and friendships. These reports include an examination between task and social relationships as well as those between group members and dyads.

Another prominent advocate of scholarship that focuses on the relationships that result from interactions, Fitch (1998) used a combination of the ethnography of communication and cultural codes theory to examine the affordances and constraints of roles, such as mothers and daughters, during actual interactions in a variety of settings. As Sandel and Ju allude to in chapter 5, Fitch (1999) was one of the first LSI scholars to examine discourse in an online discussion and make observations about the ways participants relationally positioned themselves by their claims.

Unlike the Hymes's SPEAKING categories, the analysis of relationships is an interpretive step that draws upon turn-by-turn communicative acts to formulate or draw conclusions about what the series of interactions between people imply about their relational status. By using actual interaction, a researcher can formulate hypotheses about relationships that are enacted. This is a different type of conclusion than those that are established based on self-report data (that is, interviews) about who users are to one another. For instance, one may post marital status on a Facebook profile, but then during interactions provide different status cues, such as not wearing a wedding ring.

People who interact with and through digital media often make claims about their relationships in the course of interacting, for instance by issuing a proposal such as "will you join my network (on LinkedIn)?" This relational positioning occurs as part and parcel of the process of communicating itself. That is, when we speak, we not only issue commands and directives, but we also indicate who we are to one another. Harvey Sacks's (1995) research provided an interpretive lens that illustrated the way we notice relational cues during everyday talk. For instance, Sacks refers to one sentence he heard, "the baby cried and the mommy picked it up" as a quintessential formulation of role relations that were being enacted by the speakers within this conversation. When speakers recognize that mommy and baby are part of the category of family, they are considering more of the interactional context to make sense of the statement. This same type of work is involved when we read descriptions of cell phone messages between friends, feedback to classroom peers, or talk between board members in the chapters that follow. In each of these examples (elaborated by Peters, Bouwmeester, and Sandel and Ju respectively), the relational terms reveal categories for relating: friends, class peers, and board members.

Overall, the chapters included in this section draw attention to different expectations that relational categories imply as well as the unfolding of relationships through interaction itself. They also provide some evidence for cultural implications, but we will save that focus for part III.

The next chapter serves as a clear bridge between the first section's focus on communicative actions and the way actions are part of relating.

In chapter 3, Peters describes the way members of a nonprofit board try to maintain their current relationships as well as accomplish work-related tasks by using video-conferencing software. Through Peters's deft analysis, we clearly move from task related activity to another goal that is accomplished within a similar meeting framework, and that is maintaining group identity and cohesiveness.

In chapter 4, Bouwmeester recounts evidence above assumptions (and constraints) faced by student teachers when they try to enact peer-relationships during a feedback exercise using a new digital medium. The specific tasks the software was designed to enable (rubrics) seem to hinder the ability for peers to provide the kind of informal feedback to fellow students they have been used to providing. Bouwmeester explores the ways student teachers responded to their interaction with the software as a difference between enacting a relationships with their peers as one that is more like that between a student and an instructor. Recognizing and overcoming the challenges posed by the software itself upon already existing relationships can provide information for designers to use.

A similar dialectical tension occurs in chapter 5. Sandel and Ju describe the social and professional roles performed during the use of WeChat. When students in Macau describe their use of this social networking tool, they describe their emotional reactions to posts, including embarrassment when role transgressions are made. The analysis of interactions through this digital medium helps us learn about the way typical relational roles are maintained and the new ways it might change relationships based on the availability of access to those who were formerly less able to be contacted in the social hierarchy.

Let us turn to these analyses of relationships now.

REFERENCES

Ellis, Donald G., and B. Aubry Fisher. *Small Group Decision Making: Communication and the Group Process*, 4th Edition. New York: McGraw-Hill, Inc., 1994.

Fitch, Kristine. *Speaking Relationally: Culture, Communication, and Interpersonal Connection*. New York: The Guilford Press, 1998.

Fitch, Kristine L. "Pillow Talk?" *Research on Language and Social Interaction* 32, no. 1&2 (1999): 41–50.

Keyton, Joanne. "Relational Communication in Groups." In *The Handbook of Group Communication Theory and Research*, edited by Larry R. Frey, Dennis S. Gouran, and Marshall Scott Poole, 192–222. Thousand Oaks, CA: Sage, 1999.

Murray, Janet H. *Inventing the Medium: Principles of Interaction Design as a Cultural Practice*. Cambridge, MA: The MIT Press, 2012.

Sacks, Harvey. Lecture 2(R) "The Baby Cried. The Mommy Picked It Up" (ctd). In *Lectures on Conversation*, edited by Gail Jefferson and Emanuel L. Schegloff, 223. John Wiley & Sons, 1995.

Sigman, Stuart. "Order and Continuity in Human Relationships: A Social Communication Approach to Defining 'Relationship.'" In *Social Approaches to Communication*, edited by Wendy Leeds-Hurwitz, 188–200. New York: Guilford, 1995.

THREE

"Showing We're a Team"

Acting and Relating in Online/Offline Hybrid Organizational Meetings

Katherine Peters[1]

I sit down to the September 2013 board meeting about five to ten minutes before it is scheduled to begin. I open up my computer, sign in to our video-conferencing software, and gather together the agenda and other documents that were uploaded to Google Drive earlier in the week. I pick up my mug of tea and click on Mary's name to call into the meeting. I hear a ringing tone, and then the whoosh sound as she answers. After a few seconds, the video feed pops up, and I can see her at her place at the head of the table set up with some papers and pens. To one side, I can see about half of her sister, Lise. Both of them wave and greet me as I pop up on their screen. On the other side of the screen, I see a hand attached to an arm waving, and hear the voice of Lisa, a mutual friend. I can hear Dan's laugh and Sebastian's muddled voice, but I cannot see them, and I assume that Dan is sitting next to his fiancée Lisa, which puts him close to the computer on the left and Sebastian close to the computer on the right.

As I maximize the screen, this view of one-and-a-half bodies plus a few phantom limbs on occasion takes up about two-thirds of my screen. Along the right-hand edge, I can see myself in a smaller window at the top, and soon one more image pops up as Mary adds Amanda to our call. Amanda and I wave hello; we have bonded over always being the online members in meetings. Dan, Lisa, Lise, and Sebastian shuffle the papers in front of them and wind down their pre-meeting conversations and joking as Mary calls the meeting to order. I turn to the documents in front of me, and I assume that Amanda does the same. Through the video-conferencing software I can see a total of three-and-a-half bodies, including my own, which is half of the total participants in the meeting.

Meetings, as an arguably ubiquitous feature of organizations, present an interesting dilemma when multiple organizational members must use technology to attend. This has been the case for Suicide Prevention Campaign (SPC), a small nonprofit organization in Pennsylvania. Three out of nine board members moved away from the area shortly after SPC was founded, and they have driving commutes ranging from four to twenty-two hours. Rather than aligning the busy schedules of eight twenty-somethings and a retiree to find the time to meet at least twice a year together in person, SPC has turned to technology. The board and the three organizational committees have used video-conferencing software in addition to in-person meeting spaces. The online/offline hybrid nature of these meetings has presented unique requirements, challenges, and opportunities for holding meetings as an organization. SPC and other organizations may require software designed to foster relationships between all members of the organization and to accomplish the activities that are associated with the practice of meetings

MEETINGS AND TECHNOLOGY

Meetings

Meetings have been the attention of study since Helen Schwartzman's (1989) seminal work *The Meeting*. Organizational studies scholars like Deirdre Boden (1994) have also turned their attention to this ubiquitous practice. Although meetings may on the surface seem like mundane or routine conversation, these scholars have drawn attention to the ways in which meetings are central to accomplishing the work of organizations. From the definitions provided by Schwartzman and Boden, several features are relevant to this study of organizational meetings. The three basic features of a meeting include: multiple people attending, a forewarning of the event, and a purpose sustaining the meeting. Multiple people must attend a meeting, although organizations may set more specific minimums. These participants must have made an agreement to meet, or have at least some forewarning of the event prior to its start. Finally, meetings are organized around a purpose. This sustaining purpose is usually related to the work or functioning of a larger organization. These are the most basic features shared among organizational meetings. What makes this a unique communication event is that it is where people can interact with each other to accomplish work toward a purpose that is related to organizing.

Scholars have discovered ways in which meetings are particular to different groups, organizations, and cultures. Ethnographers of communication in particular have used Schwartzman's work as the grounds for studying meetings as cultural events (Sprain and Boromisza-Habashi 2012). Studying meetings in this way may involve discovering differences between national cultures or smaller, organizational-level cultures. These differences could be between the values embedded in the meeting practice, the norms that members have for how interaction should be structured, and the expectations that they have for activities involved in meetings as well as how members should relate to each other. For example, Yamada (1990, 1992, 1997) studied some of the ways in which American and Japanese businesses conducted meetings differently. An example of this difference was that Americans value everyone's input on a topic during a meeting, whereas Japanese workers saw such a free exchange or discussion as chaotic, and instead valued the input of one or two leaders rather than all of the participants in a meeting. Americans might have viewed talk between only a few members as reinforcing existing hierarchical relationships rather than fostering egalitarian relationships between meeting members. For SPC, these large-scale cultural ideals could have affected how members act and relate to each other in meetings.

In Pan, Scollon, and Scollon's (2002) study of Chinese business meetings they found similar results. The discussion activity that Americans expected in meetings was not expected during Chinese meetings. The main action involved in Chinese meetings was the ratification of the leader's opinion, which involved the leader doing most of the talking and other members agreeing with his arguments or proposals. Similarly to Japanese meetings, this served to reinforce the hierarchical relationships between members during meetings. However, pre- and post-meeting actions were vastly different from these actions during the meeting. The authors found that Chinese workers expected decision-making, discussion, and argumentation to occur between members during pre- and post-meeting talk. Although meetings involved reinforcing hierarchical relationships through the ratification of a leader's opinion, pre- and post-meeting talk may have served to foster egalitarian relationships through decision-making, discussion, and argumentation.

Although national cultures provide us with some differences between expectations for conducting meetings on the global scene, there are also differences between more local cultures within nations or even within the same community. The particularities of culture and the differences between values, interaction structures, and expectations for acting and relating in meetings can also be seen through the differences between single organizations. Tracy and Dimock (2004) have argued that one of the important functions that meetings served was that they are "the arena in which organizational and community groups constitute who they are" (140). By establishing group identities, members were also establishing expectations for acting and relating between members, both inside and outside of meetings. This harkens back to Schwartzman's (1989) study of Midwest Community Mental Health Center. She found that the organization sought to establish itself as an "alternative organization," and thus meetings provided a unique site to practice and embody their ideals. At Midwest, staff members frequently expressed and shared emotions with each other in meetings, which served to reinforce their commitment to therapy and mental health. The group identity of Midwest was also reinforced by meetings themselves. This was shown especially through the constitution of ongoing conflicts that motivated people to continue to attend meetings, and thus continue to constitute Midwest through meeting talk. Similarly, Mirivel and Tracy (2005) found that members of a nutrition organization established a "young" and "health conscious" identity by discussing health and fitness during pre-meeting talk and having nutrition bars and water bottles for snacks.

Meetings can also serve to show differences between groups within the same organization. Ruud (1995, 2000) found that the musicians and the business operators in a regional symphony came to meetings with two different expectations of how to communicate with others in meetings. Although they differed, he argued that the codes were interdependent, as both were necessary to complete the work of the symphony organization. Baxter (1993) found differences between faculty members and administration members at university meetings where these groups needed to work with each other. She particularly found differences between what counted as official for each group. "Talking things through" was favored by faculty members, whereas "putting it in writing" was favored by members of the administration. The differences between expectations of how to communicate and what counted as official produced tensions between the groups.

Together, these previous findings have informed my examination of meetings in SPC and the cultural premises and discourses embedded within their practice. Meetings, as well as their structure and particularities, are informed by cultural preferences on both macro- and micro-levels. The macro-level that examines preferences among national cultures provides a limited view on what individuals and organizations expect meetings to help them accomplish, but do provide some framework for broader expectations. As ethnographers of communication have shown, meetings provide a place for organizations, or several groups within these, to establish and enact more particular preferences. These particular preferences for talking, acting, and relating in unique ways serve to constitute a shared group identity. Therefore, meetings are important for constituting group identities.

TECHNOLOGY AND ORGANIZATIONAL COMMUNICATION

For SPC and many other organizations, technology is an integral part of accomplishing work. Meetings, as part of work, also use various kinds of technology in order to coordinate acting and relating with other members of the organization. In organizations such as SPC, information and communication technologies (ICTs) may be integral to the ability of several members to contribute to the purpose that the meeting was called to address. For decades, organizational communication scholars have examined how ICTs, work, and users mutually influence each other (see reviews in Jackson, Poole, and Kuhn 2002; Rice and Leonardi 2013), and other scholars have been concerned more specifically with computer-mediated communication such as video-conferencing software (see Sellen 1992; Finn, Sellen, and Wilbur 1997). Many of these researchers have viewed technology as constructed by both designers and users. In organizational communication research ICTs include "the devices, applications, media, and associated hardware and software that receive and distribute, process and store, and retrieve and analyze digital information between people and machines (as information) or among people (as communication)" (Rice and Leonardi 2013, 426). ICTs can thus be used to construct unique information or communication practices, or they can be used to aid existing practices.

Although ICTs may be purchased initially or adopted to enhance or aid existing practices, over time their use can help to construct new ways of acting and relating, or change the ways that members relate to each other and act. These changes may be due to the design of communication with which the technology was created (Aakhus and Jackson 2005). When technology is based on a designer's ideals of communication, rather than the community's ideals of communication, then the technology may influence interaction between members. This in turn may influence the ways that members act and relate with each other in meetings. However, members may also use the software in ways that a designer did not intend, and thus can potentially influence the software itself and alter the design of communication. This study seeks to explore the ways that members actually use the software during meetings to provide evidence of how members may use the software in ways the designer did not intend. Over time, such use could influence the software itself and alter the design of communication. It is in this way that designers and users together create and influence technology. The ways that software influences how people interact with each other could have implications for the identity and cultural expectations of organizational members. The design of technology used during meetings is therefore consequential to the conduct of organizations and their ability to constitute a desired shared identity.

Researchers of human-computer interaction (HCI) have investigated the use of software for various purposes, including holding online meetings. Aside from one of the obvious considerations for selecting one kind of software over another, price, members must also consider whether to choose audio-only or audio and video technology such as video-conferencing software. A benefit of video-conferencing software is that it does allow for more nonverbal cues from communicators, which can help to facilitate process and content coordination between members (Whittaker and O'Conaill 1997). Although enhancements to technology are made over time, studies read years after they are conducted can still provide valuable insights to designers. For example, the basic kinds of video-conferencing software have not changed, although features differ from one product to another. Picture-in-picture software combines each participant's video feed in different sections on a screen, and broadcasts generally the same screen to all participants (Sellen 1992, Finn 1997). Audio feeds are usually also combined, which may make differentiating different people, voices, and conversations difficult (Sellen 1992, Finn 1997). Picture-in-picture is the general type of software that SPC has used in meetings. Although picture-in-picture software allows a viewer to see all of the people interacting at once, and their nonverbal cues, picture-in-picture video-conferencing may not facilitate certain aspects of interaction. For example, Sellen found that users were unable to establish eye contact, make side comments to other participants, hold parallel conversations, or selectively listen to different participants talking at once (Sellen 1992). Some of these disadvantages have been addressed by additional features such as text chat, but some of these same limitations still apply depending on the design of the software.

Researchers have examined the differences between same room conversations and video-conferencing set-ups, and how these relate to the values that participants had for the practice they were holding with or without the video-conferencing set-up. Sellen (1992) found that same room conversations tended to be rated as "more interactive" by participants than video-conferencing conversations. This might be due to a difference in the values that humans bring to an interaction, and which of these values are enabled or constrained by the technology's design. Sellen, Rogers, Harper, and Rodden (2009) advocated that "human values, in all their diversity, should be charted in relation to how they are supported, augmented, or constrained by technological developments" (63). This is the purpose of this study, to uncover the specific values and beliefs that participants bring into a practice, rather than relying on more universal human values. The ethnography of communication and cultural discourse analysis provide a way to uncover these particular values and beliefs shared among organizational members, and how the technology supports or alters them. Together, these bodies of research on meetings and ICTs have prompted me to ask: How are discourses of acting and relating supported, altered, or constrained by technology in online/offline hybrid organizational meetings?

METHODS

This chapter is based on two years of ethnographic research including participant observation, interviews, and document collection with Suicide Prevention Campaign (SPC) between 2012 and 2014. SPC is a small non-profit organization in central Pennsylvania, and it was founded by Mary in 2010. SPC's mission is to provide education and resources about suicide prevention and mental health awareness to teenagers, teachers, staff, and community members. In 2012, Mary created the board of directors by inviting many of the people who had supported her in the two years that she started and ran SPC on her own. These volunteer board members included her sister (Lise), one of her former high school teachers (Doug), and several friends (Dan, Lisa, Sebastian, and me). During my research, two additional board members, Amanda and Theresa, were added.

I joined SPC's board four months before I started my research with them. During my observation period, I attended ten out of fourteen meetings held by SPC, which includes all of the board meetings held during that time. Meetings were scheduled when Mary or a committee chair determined that they were needed, rather than holding regularly scheduled meetings. The only regularly scheduled meeting was the annual board meeting in December each year. For two months all of the members volunteering on the board could attend face-to-face meetings and did not need technology. After this, Amanda and I moved away from the area to attend school and could not feasibly return for meetings as they were scheduled. Mary continued to hold face-to-face meetings around her six-foot kitchen table for those who lived nearby, and those who did not were invited to attend meetings using a video-conferencing software that I refer to as Voom. Amanda, Theresa, and I regularly attended meetings through Voom only, and Sebastian occasionally participated in meetings this way as well. Voom is available to download for free and was found and chosen by Mary, who performed a Google search for free group video-calling software. In Voom, calls must be accepted by a receiver like a telephone call, so online members could not automatically join the conversation. The software company lists values of "personal connection" that they want to create through their software, so Voom was not necessarily intended for use during organizational meetings, or even those of an online/offline hybrid nature like the ones I study with SPC.

I audio-recorded each meeting using a second software program, and then constructed transcripts and field notes for each. Throughout this time, I also conducted interviews with three members of the board of directors about meetings. Together, this produced 541 minutes of recordings and 536 pages of transcripts and field notes. I also collected documents and participated in an online messaging board and email system that members often used instead of or between meetings to supplement the data that I gathered from participant observation in meetings and interviews.

After gathering this data, I used analytic methods informed by the ethnography of communication (Hymes 1974) broadly, and cultural discourse analysis (Carbaugh 2007) in particular, to interpret which values were embedded in these online/offline hybrid organizational meetings. I first used Schwartzman's (1989) adapted framework of Hymes's (1972) SPEAKING model to separate analytically the components and aspects of meetings. Her framework involves separating data about meetings using the following analytic concepts: participants, channels and codes, frame, meeting talk (which includes the sub-components of topic and results, norms of speaking and interaction, oratorical genres and styles, and interest and participation), norms of interpretation, goals and outcomes, and meeting cycles and patterns. This framework helped me to develop a general sense of interactional patterns in meetings and what occurred, and then I used cultural discourse analysis (Carbaugh 2007) to reveal more specific information about values of acting and relating during meetings.

By adding cultural discourse analysis categories, I found that the intersections between acting and relating were the most interesting when regarding the online/offline hybrid nature of meetings in SPC. Cultural discourses of acting deal with what members should be doing or take themselves to be doing (Carbaugh 2005). These are the actions that members explicitly state that they are doing and also what practices they engage in that might not be so explicitly stated (Carbaugh 2007). To find and restate these cultural premises of acting, I began to examine the actions members explicitly told me were involved in meetings. Then I determined the frequency of these actions in practice as well as how these explicated actions set up expectations that were both enacted and enabled or constrained by the technology.

These findings led me to analyze cultural discourses of relating, by investigating messages that illustrate how members relate to each other during their meetings (Carbaugh 2005). These relations may be presumed prior to an interaction and also constituted within interaction (Carbaugh 2007). In order to uncover cultural premises of relating, I examined both how members referred to each other in interviews or pre-meeting talk and also how they referred and related to each other during meetings.

The ways that members related to each other were based on particular ways of acting that they associated with certain kinds of relationships. I again also examined how these expectations of relating were enabled or constrained by the technology.

CULTURAL DISCOURSES OF MEETINGS

Cultural Discourses of Acting

When Mary or a committee chair called a meeting, they and other members of SPC expected certain actions to take place. If these actions were impeded or missing, then a meeting could have been considered less productive or successful. During my interviews, members identified two different categories of acting: informing and "showing we're a team." Informing actions gave structure to the meeting, and this structure provided a norm for how meetings should be run. These actions are outlined on an agenda that is distributed one to two weeks before a meeting. When talk turned away from the structure, Mary, as the usual meeting chair, called out the discrepancy and noted the agenda item to which the talk might relate. In an interview with Mary, she gave an account of the kinds of informing actions that comprised this structure.

Excerpt 1: Interview with Mary (lines 427–432)

M: During that meeting we usually go over um minutes of the last meeting, we discuss you know current projects that we're working on, some of our future goals. Um for the annual meeting we will discuss the year in review everything we've done within this past year, and our projected accomplishments and finances and stuff for the upcoming year as well, so we kind of we do a review and a current and uh future year plan going on for our board meetings at our annual meeting.

The actions that Mary described for board meetings, and the annual board meeting in particular, largely had to do with informing other members about the past, present, and future. In this excerpt, Mary provides an account of the sequence of actions that give SPC's meetings structure. First, members "go over" the last meeting's minutes, thus reviewing a record of the previous meeting's talk and topics. Members then "discuss" current projects and the year in review, what has happened within the last year, goals that were currently being worked on, and the current financial situation. Finally, members may also "plan" for the following year. This involved stating "projected accomplishments" and a budget. Thus, meetings both articulated and formed SPC's past, present, and future. These three times, and the actions associated with them, coexisted through the informing actions of "going over," "discussing," and "planning" in which members engage. The structure of meetings in SPC tended to take a linear path through time, first "going over" what had happened, then "discussing" present projects, and finally "planning" for the following months or year. Mary, as both president and chair of the board, tended to lead and dominate each action progressing through the times that were articulated during meetings. Voom generally supported these informing actions, as one member was speaking at a time and it was relatively easy to discern who was talking at once.

When a meeting was called, the overt purposes of the meeting also tended to be informational. Meeting agendas listed topics as either a report by an individual or as an item requiring joint discussion. However, even before a meeting agenda was drawn up, the committee chair or president first had to decide if a meeting was warranted. This involved determining whether a meeting was needed to accomplish the work that needed to be completed, or if the work could be coordinated or completed through online-only means such as the message board or email systems. In an online interview, Lise described three circumstances in which a meeting would be called.

Excerpt 2: Interview with Lise (lines 52–58)

Le: In SPC, we usually have formal meetings when either 1. something very important needs [to be] discussed, 2. when we feel like there needs to be a touch-base (for example: if we haven't had an in-person meeting in many months and we feel that people are losing touch of what's going on), or 3. for formalities like the annual board meeting.

The first and third circumstances that she mentions are related to informing actions. The first circumstance that she mentioned was when "something very important needs [to be] discussed." In the meetings I observed, these very important topics had been to discuss the potential acquisition of a smaller nonprofit organization, to elect a new treasurer, and to plan for some larger events. In the case that a meeting was called to discuss "something very important," then the meeting talk tended to revolve around that topic, and little else. Meetings called to address these topics tended to abbreviate other parts of the structure, or relate them directly to the issue rather than the broader organization. The third circumstance she described was similar to this first one in that the annual board meeting was called for informational purposes. The annual board meeting was held to meet requirements for their 501(c)(3) nonprofit status, and to fulfill the purposes of reviewing the fiscal year, planning for the next year, and re-electing board members. Meetings scheduled for this reason closely follow the structure Mary described in the first excerpt above. These two reasons required features of the software that would aid in informing-sharing actions, where one person may speak for an extended amount of time and the other members would need to hear them clearly. Meetings of these kinds would also require features that enable discussion in a larger group. Voom supported these actions by providing the ability for both online and face-to-face members to contribute to the conversation, and any member could hold the floor for the lengthy turns that were required in these circumstances.

The second circumstance Lise described was more unique, and sometimes it was used in conjunction with the first as reasons for a meeting. A meeting in September 2013 was called to discuss and generate some shared guidelines and norms around meetings and committees and also to "touch-base" with the board, who had not met in nine months. The guidelines and norms around meetings and committees were the main topic of the meeting that Mary wanted to discuss. However, she framed the meeting as a sort of "touch-base." The following excerpt is from the beginning of that board meeting when Mary framed the reasons for calling the meeting and what she wanted to talk about first.

Excerpt 3: September 2013 board meeting (lines 43–47)

M: Ok perfect so this is not going to be super long it's gonna be a really quick meeting but I really felt like we just need to get together and just kind of hang out and have some contact with each other and go over pretty much what we've been doing this year and what and for the rest of the year and goals and all that fun business kind of stuff.

Mary's account for calling and holding this "really quick" meeting was because she felt like the board of directors needed to "get together," "hang out," and "have some contact with each other." These actions had to do less with the informing actions of meetings, and more to do with actions that may have helped to "show we're a team." They also provided some informality to meetings, as "hanging out" was an action that would also describe what these people would do as friends even if there were no meeting. These "showing we're a team" actions existed alongside the informing actions during this meeting, such as "going over" the past and planning for the future. "Showing we're a team" was a category of acting that involves several kinds of actions and outcomes of meetings in addition to the three actions Mary described at the start of this meeting. Mary described some more of these actions and outcomes during her interview, which are in the following excerpt.

Excerpt 4: Interview with Mary (lines 449–453)

M: I encourage people to have fun especially since this is all volunteer
      based, even the board of directors is volunteer based, I don't want
      Suicide Prevention Campaign to ever be a burden on anyone else so
      that they feel they have to get all dressed up for a meeting or that they
      have to conduct themselves professionally um just to you know throw
      around ideas at each other and to-to discuss things and to make an
      impact.

In contrast to excerpt 2 above, where Lise was describing the purposes
of "formal meetings," Mary seemed to be describing the informal aspects
of meetings. She said that she encouraged people to "have fun," rather
than getting "all dressed up" and having to "conduct themselves professionally." Frequently members of SPC joked around with each other and
attended meetings in t-shirts and jeans. In addition to "showing we're a
team" actions being more informal, members also expected meetings to
include more informal ways of sharing and discussing information, such
as "throwing around ideas." The informality of these actions may have
served to emphasize the more friendly relations between members of
SPC, and to reinforce the informality of meetings. The norm for meeting
purposes was that they could be called for a purpose usually related to
informing, but may also be called to "show we're a team." The outcomes
of meetings also seem to include these patterns, with possibilities of creating a sense of having had "fun" together in addition to articulating and
forming the past, present, and future of SPC. If these were the expectations that members have for acting, how were these enacted in online/
offline hybrid organizational meetings?
    One meeting demonstrated some of the differences in the ways that the
software affected how members acted. The December 2013 board meeting started with the introduction of a guest before the meeting. Theresa
was at that time a potential board member who was invited to introduce
herself to the board and to answer any questions that members had before
they voted on whether or not to elect her as a director. Theresa's presence
as a guest might have prompted Mary to make explicit statements about
how the software affects meeting interaction, but these effects have been
restated at the beginning of several meetings since then. The following
excerpt starts after about five minutes of pre-meeting talk, and just prior
to this excerpt Mary had asked meeting participants what the exact time
was so that this could be noted in the minutes to open the meeting. Saying
and including the exact time that the meeting begins is part of the formal
opening of the meeting frame, so this excerpt marks the beginning of the
meeting.

Excerpt 5: December 2013 annual board meeting (lines 29–38)

M: Alright 2 oh 9 we're officially starting Amanda is taking minutes for
       this meeting now our people on the phone and on the computer it is
       kind of hard for them to hear us it's hard for them to hear each other
       and it's hard for us to hear them so um. . . .
La: Sweet
M: There's lots um it's hard to hear everyone so just make sure you guys
       are speaking loud and clearly especially if you're on technology yell
       at us.

Two norms for interaction developed around SPC's use of video-
conferencing technology. Within the face-to-face meeting space in Mary's
kitchen, the laptop was set up at one end of a six-foot table, so those who
sat farthest from the computer were unable to hear and be heard by those
meeting online. For this reason, the norms of interaction included "speaking loud and clearly" or "yelling" in order to ensure everyone could hear.
Another reason that these norms of interaction developed was due to the
slightly asynchronous feed of the software, which manifested in a delay
in online members hearing face-to-face members' talk, and vice versa. The
outcome of these norms and the design of the technology was that people
who met face-to-face in Mary's kitchen tended to speak more frequently
during meetings. In many meetings, online members had to talk loudly
to try to be heard over the cacophony of talk in the in-person space. To
try to address this discrepancy in the last two meetings that I observed,
online members started to physically raise and wave their hands in order
to signal for a turn and get the attention of the chair. Members meeting
face-to-face continued to use more ordinary turn-taking cues in order to
gain control of the floor (Asmuß and Svennevig 2009; Sacks 1992). This
new turn-signaling device combined with the slight asynchronicity in the
software resulted in turns with boundaries of a few seconds preceding
and following an online member's contribution, as opposed to the turn-
overlap that occurred before. Online members' turns are therefore more
obviously recognized than face-to-face members' turns.

The video-conferencing software compounded some of the benefits and
drawbacks of the online space by affecting the visual aspects of meeting as
well. This software was picture-in-picture, so when multiple people were
using the online space to meet, each person's screen was made smaller
in order for everyone, including the user himself or herself, to fit on one
screen. If the program was maximized on a computer screen, then one
person's feed could be made larger than the others. Typically, I used this
feature to make the largest screen Mary's feed because her room included
the largest number of participants. Visually, this emphasized the physical
space and the actions happening in that video feed. This set-up made the

other online participants much smaller by comparison. The laptop on the kitchen table was set up so that the program divided the screen equally between each of the attendees' feeds. No online member was emphasized over any other, but all of the online members' faces were proportionally dwarfed in comparison to those meeting face-to-face. Other participants made individual choices as to which of these two views to use.

The video-conferencing software minimally affected the ways that members informed each other, through "going over" or "discussing." Mary tended to talk the most in meetings, as president it was her role to report on the past, present, and future actions of the organization. However, when it came to "showing we're a team," online members were at a disadvantage. These informal actions described as "having fun" and "hanging out" often took place during pre- and post-meeting talk, as well as side conversations between members meeting face-to-face. Face-to-face members were invited directly after the meeting was concluded to stay and socialize while finishing the food Mary ordered for the meeting, which online members could not do. Face-to-face members often talked with each other until the online members called in, and then online members were occasionally included. For example, Mary once called me five minutes earlier than other members because she wanted to chat before other members arrived. However, online members usually called into the meeting only a few minutes prior to the start time, and signed out as soon as the meeting concluded. According to Mirivel and Tracy (2005), this pre-meeting talk, and perhaps also post-meeting talk, may have served to develop SPC's organizational identity. During meetings these informal actions could be observed by those of us who were not physically present because we could hear laughter and quieter side conversations occurring between members. Online members could not easily participate in these side conversations because the software was slightly asynchronous and online members would also have to yell in order to participate. The video-conferencing software provided relatively straightforward features that allowed members of SPC to accomplish informing actions. However, the software made it more difficult to participate in side conversations about topics not included on the agenda or chatting with others before and after the meeting. Thus, this made it more difficult for online members to engage in the organizational values of "showing we're a team" and "having fun" together, making these relevant only for those meeting face-to-face.

CULTURAL DISCOURSES OF RELATING

Further insight that would provide information for designers also could be found through how technology affected the ways members related to each other during online/offline hybrid meetings. Members of SPC have expressed wanting to be a team together although they have an official organizational hierarchy. The organizational hierarchy was presumed prior to meeting interaction, as were informal and friendly relations between members. An organizational chart details the hierarchical relations between members. The board of directors has more authority than the organization's officers who were the secretary, a role fulfilled by Lisa; the treasurer, a role fulfilled by Sebastian; and the president, a role fulfilled by Mary. However, these officers were also part of the board of directors who oversaw the performance of the officers. Committee members who were not on the board of directors had the least amount of power according to the organization's chart, but the committees were mostly comprised of board members.

Beyond the formal hierarchical relationships between SPC's member roles, most members had been friends before volunteering for SPC and its board. These outside relationships may have contributed to the preference to relate as a "team." This was made explicit during interview comments. Consider Lisa's description of how members interacted through a mix of fun and work. Prior to this excerpt, she described meetings as including a mixture of "fun communication" and "business work." She then elaborated on these categories later in the interview when I asked her to explain what she meant.

Excerpt 6: Interview with Lisa (lines 82–85)

La: We can generally joke around with each other because we know each other and know where we are in life. We can joke with you about your education and the never-ending homework load, but two minutes later we can also decide who is going to fill what role at our next educational event.

Lisa clarified that "fun communication" involved "joking around," whereas "business work" involved "deciding" something about the work that SPC was doing. Holmes and her colleagues (Burns, Marra, and Holmes 2001; Holmes 2000; Holmes 2008) found that humor and small talk were more likely to occur with meetings containing a majority of female participants and that this small talk and humor created more personal relations among members. Although most of the members of SPC and the board of directors were female, in this organization "joking around" with each other seems to hinge on "knowing each other" rather than the gender of participants. Staske (2002) described the way conversational participants' knowledge about one another can index a close relationship between the pair where such knowledge is expected. In the case of SPC board meetings, when Lisa was referencing knowledge of "where [participants] are in life," she also was indexing close relationships or friendships between members of SPC. She even provided an example for me, describing my own educational pursuit and my "never-ending home-work load." Another frequent kind of joking occurred between Mary and Dan, who sarcastically referred to each other as "arch-enemies" based on a history of not agreeing with each other and frequently debating. This inside joke often was mentioned when Dan disagreed with Mary on a small decision. "Joking around," therefore, was used by members who had knowledge about each other to relate to each other personally. "Joking around" might have been easier for face-to-face members to enact because they could make a quick, overlapping comment to the person sitting near them, rather than interrupting the flow of conversation by raising one's hand and waiting to be called on in order for a joke to be heard by other members.

In contrast, "business work" included delineating which "role" participants would play in the next work event. Each event "role" included both a set of tasks to be completed as well as an implied reporting relationship. These reporting relationships indicating dependencies and accountabilities often determined how members would relate to each other during the event. When discussing these kinds of work roles, participants who might otherwise relate to one another as friends, immediately switched into role-related interactional styles to accomplish work together. As I mentioned above, the organizational hierarchy affected their roles so that they include certain accountabilities and structure.

How did the meeting software influence the way "business work" was enacted in the meetings? As an example, in the board meeting in September 2013, Mary brought up the topic of meetings in order to discuss and agree as a group on the frequency and structure for the future. This topic was related to a prior joke from Lisa who suggested having "fiestas" with Mexican food and a talking sombrero to indicate speaking turns. She made this joke as a suggestion for how to make sure meetings were not boring. Mary started talk on the topic of meetings by stating her expectations, which may have also reflected the organization's expectations about meetings.

Excerpt 7: September 2013 board meeting (part 4, 288–309)

M: The first thing I want to talk about here um I do want to talk about our meetings a little bit. Um cuz the biggest fight that I have with myself with meetings is I want to be able to meet with people to, you know, show that we're a team so everyone gets together, and we get to brainstorm together, we get to talk face-to-face or you know webcam or whatever it is. But I also know that everyone has very busy lives, and I don't want to have meetings just for the sake of having a meeting. Because we do everything online, I can get information to you guys in an instant and not have to schedule a meeting. But I still want to be in communication with you guys. ((laughing)) And like have that face-to-face interaction. Because I want you guys to know that we are a team and that we're all working together. So that's kind of my biggest struggle and I just want to know some of your thoughts on it. What do you guys prefer, do you hate having meetings? Ha- are we having enough meetings? Should we have less meetings? Should we not call them meetings? Should we call them fiestas? Would that make it better?

La: Probably, yeah probably
((laughter))
M: Probably. Would you come to a SPC fiesta?
La: Who wouldn't come to a [fiesta?
Le: [I feel like they're more productive. I'm more
                        productive.
S: Yeah there's still something about (())
Le: Yeah then you're focused and when you're online not everybody's there at the same time and you're meeting at different times, you don't know what you're speaking about.

Mary's monologue began with a characterization of a "fight" that she was having with herself. She wanted to hold meetings because they "show we're a team," which was a certain kind of action that allowed members to relate to each other in a particular way. In this characterization of meetings, there was a connection to relational communication and being a whole together in meetings, rather than scattered parts. Meetings allowed "face-to-face interaction" and "being in communication" with other members of SPC. As Katriel and Philipsen (1981) researched among U.S. English speakers, "communication" was a characterization of talk that usually was referenced and used in personal relationships. "Being in communication," through meetings, was a way for the members of SPC to relate to each other on a more personal level. However, at the same time, members of SPC were also relating to each other in organizational roles on a "team" in addition to the personal level of being a "team." Being a "team," and relating to each other to "show we're a team" involved the actions of "getting together," "brainstorming together," "talking face-to-face," and "working together." Relating to each other as a team involved doing work together, rather than separately, which meant that meetings visually reemphasized the "team" through being able to see and communicate with multiple people at once who were focused on the same matters. Voom allowed online members, many of whom lived too far away to attend meetings physically, to have the chance to participate in these actions during meetings.

The visual aspects of Voom that members used provided some benefits and some drawbacks toward relating to each other as a "team." The video-conferencing software allowed all online members to be seen by in-person members, as opposed to a conference call on a telephone or similar audio-only channel. Online members could also see each other and some of the face-to-face members, which aided with recognizing nonverbal cues and also visually representing the "team." However, the online members could typically only see Mary and perhaps half of one or two other members' bodies. Both the camera and the software seemed to privilege one-on-one interaction, with the camera angle and view only allowing for one or two bodies in a screen at any one time. Therefore, online members could not see a full representation of the "team" that was meeting. Instead, the view provided by this video-conferencing set-up emphasized a one-on-one and hierarchical relationship between Mary and the online members. This emphasized relating to each other in organizational roles, rather than more informally as friends, and affected the online members more than those who often met face-to-face.

At the end of Mary's turn in the above excerpt, and the following three turns, there was an example of "fun communication" through joking around with each other. Lisa's suggestion of a fiesta from earlier in the meeting resurfaced here and was posed as an option to improve meetings. Although joking may serve multiple functions such as relieving tension or differentiating roles (Tracy, Myers, and Scott 2006), for SPC "joking around" seems to be part of the "fun communication" that Lisa had described above, and thus fosters personal relations between members. Lisa's suggestion indicated that she, and perhaps also other members, preferred meetings that are more informal in nature. Informal meetings would include kinds of interaction, such as joking or side conversation, that foster this informality, and the software used to meet would ideally include features to enable this kind of communication between all members. In practice, side conversations and joking were hard for online members to participate in during meetings. As a researcher, I found that it was hard to hear and to transcribe side conversations in meetings, let alone to participate in them as an online participant during meetings. Not only were online members prevented from participating in side conversations due to the slightly asynchronous feed, but online members also could not choose to attend to these side conversations over whatever audio was loudest on the audio feed.

The joking between Mary and Lisa from the excerpt stood in contrast to the "business work" that continued when Mary's sister Lise stated her opinion on meetings, calling them "more productive" than online communication. The use of "they" could be interpreted to be referencing "fiestas" here, but because the only "fiesta" was held three months after this discussion and was still referred to as a meeting, Lise would only have been able to talk about meetings themselves. Lise further stated that meetings were valued for synchronous communication and staying on the same topic so that people knew what they were "speaking about." This is in opposition to "online," which refers to the online message board and email service that members of SPC used in between meetings. They started to use this service because committees and the board might go for months between meetings due to availability and task demands. Meetings, even with the use of Voom, are valued because they provide a collective space and attention to matters immediately, rather than waiting for people to have time to check, read, and respond on the social media platform. Unlike posts on the message board system or emails, meetings could create a sense of being a "team" that the other kinds of communication lacked.

DISCUSSION AND CONCLUSION

Throughout this chapter I have described SPC members' expectations of holding meetings, and also how members conducted themselves during meetings. These expectations and interactions involved cultural discourses of acting and relating in online/offline hybrid organizational meetings. Members expressed and used norms of acting to guide what should or should not be done in meetings. Members also expressed and used norms of relating to guide how members should and should not relate to each other in meetings, and to define their relationships with each other during these situations. However, the video-conferencing technology used for some members to meet was not simply a neutral channel through which members could interact in expected ways. The video-conferencing software affected some of the norms that were created regarding meetings, thus enabling and constraining valued ways of acting and relating.

To conclude this analysis I want to summarize and reemphasize the ways that acting and relating differed for online-only and in-person meeting members. This summarizes the ways in which technology designers might construct personae using the ethnography of communication and cultural discourse analysis in order to use this as a resource for design. The expectations that members held for acting and relating in meetings provide designers with the expectations that they also would have for the technology they use to meet with fellow coworkers. Although these expectations are particular to an organization or group, features designed from a persona of this kind could allow other organizations or groups to access these features in different ways that could help them to better integrate technology into other online/offline hybrid communication practices. I now restate the expectations that members had for acting and relating, how their interactions aligned with these espoused expectations, and also how the technology both enabled and constrained interaction from aligning with these expectations.

Members of SPC expected actions in meetings that aligned with informing others or "showing we're a team." Acting to inform primarily involved relating to others in meetings as coworkers. These informing actions included "discussing," "deciding," and "going over" information, among other actions. Informing tended to include one member speaking for lengthy periods of time, sometimes with the aid of documents that were included with the meeting agenda, and then some comments and perhaps voting from other members. These indicated and reinforced hierarchical and organizational relationships. Mary, the president and chairwoman of the board, primarily directed these actions, if she did not complete them herself. These actions could also be characterized as "business work," as Lisa mentioned in our interview. Informing others was prevalent in the interaction among members during meetings. The topical focus of meetings was on information or work that could be accomplished for SPC. This related members to each other as coworkers focused on goals and outcomes for the organization.

Acting to inform and relating as coworkers were expectations that were mostly supported by Voom, the video-conferencing software members used. When online members wanted a turn, as long as they spoke loudly, they were granted their turn, and in many cases even if this was an interruption. Due to the slight asynchronous delay in the feed, there was usually a pause before an online member restarted or repeated their turn, and a pause after his or her turn for the face-to-face members to receive what was said. This same buffer was not granted to face-to-face members, who were subject to more informal turn-taking cues and patterns. Online members could also hear the informing contributions of the face-to-face members through the combined audio feed. The visual aspects of the software similarly supported informing and relating as coworkers. The in-person space usually took up most of the screen, with online members taking up smaller screen spaces along the right-hand side of the screen. This visually emphasized the in-person meeting space, where Mary and typically one or two half-bodies could be seen. Mary as the president and chairwoman of the board usually chaired meetings and spoke the most during meetings. She was visually emphasized through the software, and the software helped to represent her role. The view provided to online-only participants missed some of the gestures of other participants and made determining eye contact and gaze difficult.

Members also expected to relate to each other as a "team" in meetings, and for meetings to include actions that "show we're a team." This was a key term used to establish a group identity for SPC, and meetings provided the time and place for members to construct this identity together. Accomplishing work "together" seemed to be an important part of relating as a team, as well as including "fun communication" between members. This involved "joking" with each other, which portrayed the close relationships that some members had with others, such as friendships or familial relationships. "Getting together" and "hanging out" were actions that members expected to accomplish by holding meetings with each other as well. These also indicated more informal ways of relating with each other, perhaps as friends or as a "team." Much of the "fun communication" and other actions that allowed members to relate to each other more informally involved pre- or post-meeting talk as well as side conversations.

Relating as and acting to show a "team" were limited in support from Voom. Face-to-face members could visually see the entire "team" represented in front of them. Online members' pictures were smaller in size, and thus were perhaps less visually emphasized than face-to-face members during meetings. Online members, however, could not see all of the participants at meetings. They could see the other members meeting online, as well as Mary and maybe one or two half-bodies on screen. Online members were also limited in their ability to participate in pre-meeting or post-meeting conversations with other members. They were typically not called until just before the meeting had started and other face-to-face members had already arrived. Face-to-face members talked with each other, and sometimes interacted with the online members as well, but these conversations between online and face-to-face members were affected by the asynchronous feed. Finally, online members could not participate in side conversations held during the meeting, nor could they attend to them, which supports Sellen's (1992) assertion about picture-in-picture video-conferencing software. This prevented online members from fully participating in "showing we're a team" and enacting more informal relationships with other members, especially those meeting face-to-face. Voom also did not provide any additional functions, such as text chat, to make side comments to other participants.

Since designers do not solely determine how technology will be used it is important to consider particular use-cases. In the case of SPC using Voom for some members to meet with other face-to-face members, the software best supported one-on-one interaction to inform. In an example provided above, Mary was able to inform all participants, including those online-only and face-to-face, about the expectations she and SPC held regarding meetings. The software also facilitated communication between online and face-to-face members, with some adaptation from online members to raise their hands to indicate wanting to take a turn.

More informal communication, such as pre- and post-meeting talk and side conversations, as well as other expected ways of acting and relating in meetings were not well-facilitated by this software. In SPC, members expected to relate informally to each other as friends or as a "team," by "joking around." There were no functions provided by this software to aid in these actions, and the slight asynchronicity affected online members' abilities to participate in fast-paced joking or side conversations. The software also did not provide a wide view to include all of the face-to-face members, so online members did not get the same effect of being a part of a team. Simply, Voom did not include features that could help SPC construct the group identity of "team" that they desired to create during meetings. These drawbacks are some of the reasons why SPC did choose to stop using Voom at the end of my observation period and started searching for another free video-conferencing software instead.

By conducting cultural discourse analysis about the way specific interactions transpire as well as their interpretation by participants through extended interviews, we can offer suggestions for ways to improve the software design. Such insight could result in more intuitive and user-friendly designs that help to facilitate ways of acting and relating to each other that are desired by users. In this case, Voom could have included different view possibilities or a text chat feature to facilitate participants' side conversations and more informal ways of acting to aid in relating to other members as a team. Improvements made based on this kind of research could make users feel more satisfied with the software, and thus perhaps more invested in the program over other possibilities.

NOTE

1. Publisher's Note: The interviews used as supplemental research in this text were all conducted with the participants' knowledge and agreement that these interviews would be used in a later publication.

REFERENCES

Aakhus, Mark, and Sally Jackson. "Technology, Design, and Interaction." In *Handbook of Language and Social Interaction*, edited by Kristine L. Fitch and Robert E. Sanders, 411–36. Mahwah, NJ: Erlbaum, 2005.

Asmuß, Birte, and Jan Svennevig. "Meeting Talk An Introduction." *Journal of Business Communication* 46, 1 (2009): 3–22. doi:10.1177/0021943608326761.

Baxter, Leslie. "'Talking Things Through' and 'Putting It in Writing': Two Codes of Communication in an Academic Institution." *Journal of Applied Communication Research* 21 (1993): 313–26. doi:10.1080/00909889309365376.

Boden, Deirdre. *The Business of Talk: Organizations in Action.* Cambridge, UK: Polity Press, 1994.

Burns, Louise, Meredith Marra, and Janet Holmes. "Women's Humour in the Workplace: A Quantitative Analysis." *Australian Journal of Communication* 28 1 (2001): 83–108. Accessed February 6, 2015. http://search.informit.com.au/documentSummary;dn=200112196;res=IELAPA.

Carbaugh, Donal, *Cultures in Conversation.* Mahwah, NJ: Lawrence Erlbaum Associates, Inc., 2005.

Carbaugh, Donal. "Cultural Discourse Analysis: Communication Practices and Intercultural Encounters." *Journal of Intercultural Communication Research* 36, no. 3 (2007): 167–182. doi:10.1080/17475750701737090.

Finn, Kathleen E. "Introduction: An Overview of Video-mediated Communication Literature." In *Video-mediated Communication,* edited by Kathleen E. Finn, Abigail J. Sellen, and Sylvia B. Wilbur, 3–21. Mahwah, NJ: Lawrence Erlbaum Associates, 1997.

Holmes, Janet. "Politeness, Power, and Provocation: How Humor Functions." *Discourse Studies* 2 (2000): 159–185. doi:10.1177/1461445600002002002.

Holmes, Janet. *Gendered Talk at Work: Constructing Gender Identity through Workplace Discourse.* Malden, MA: Blackwell, 2008.

Hymes, Dell. "Models for the Interaction of Language and Social Life." In *Directions in Sociolinguistics: The Ethnography of Communication,* edited by John Gumperz and Dell Hymes, 35–71. New York: Basil Blackwell Inc., 1972.

Hymes, Dell. *Foundations in Sociolinguistics: An Ethnographic Approach.* Philadelphia, PA: University of Pennsylvania Press, 1974.

Jackson, Michele H., Marshall S. Poole, and Timothy Kuhn. "The Social Construction of Technology in Studies of the Workplace." In *Handbook of New Media,* edited by Leah A. Lievrouw and Sonia Livingstone, 236–253. Thousand Oaks, CA: Sage, 2002.

Katriel, Tamar, and Gerry Philipsen. "'What We Need Is Communication': 'Communication' as a Cultural Category in Some American Speech." *Communications Monographs* 48 (1981): 301–17. doi:10.1080/03637758109376064.

Mirivel, Julien C., and Karen Tracy. "Premeeting Talk: An Organizationally Crucial Form of Talk." *Research on Language and Social Interaction* 38 (2005): 1–34. doi:10.1207/s15327973rlsi3801_1.

Pan, Yuling, Suzanne W. Scollon, and Ron Scollon. *Professional Communication in International Settings.* Malden, MA: Blackwell, 2002.

Rice, Ronald E., and Paul M. Leonardi. "Information and Communication Technologies in Organizations." In *Sage Handbook of Organizational Communication,* edited by Linda Putnam and Dennis Mumby, 3rd edition, 425–488. Thousand Oaks, CA: Sage, 2013.

Ruud, Gary. "The Symbolic Construction of Organizational Identities and Community in a Regional Symphony." *Communication Studies* 46, no. 3-4 (1995): 201–221. doi:10.1080/10510979509368452.

Ruud, Gary. "The Symphony: Organizational Discourse and the Symbolic Tensions between Artistic and Business Ideologies." *Journal of Applied Communication Research* 28, no. 2 (2000): 117–143. doi:10.1080/00909880009365559.

Sacks, Harvey. *Lectures in Conversation,* volumes 1–2. Malden, MA: Blackwell, 1992.

Schwartzman, Helen B. *The Meeting: Gatherings in Organizations and Communities.* New York, NY: Plenum Press, 1989.

Sellen, Abigail J. "Speech Patterns in Video-mediated Conversations." In *CHI '92 Proceedings of the SIGCHI Conference on Human Factors in Computing Systems.* New York: ACM, 1992. doi:10.1145/142750.142756.

Sellen, Abigail, Yvonne Rogers, Richard Harper, and Tom Rodden. "Reflecting Human Values in the Digital Age." *Communications of the ACM* 52, no. 3 (2009): 58–66. doi:10.1145/1467247.1467265.

Sprain, Leah, and David Boromisza-Habashi. "Meetings: A Cultural Perspective." *Journal of Multicultural Discourses* 7, no. 2 (2012): 179–189. doi:10.1080/1744714 3.2012.685743.

Staske, Shirley. "Claiming Individualized Knowledge of a Conversational Partner." *Research on Language & Social Interaction* 35, no. 3 (2002): 249–276. doi:10.1207/S15327973RLSI3503_1.

Tracy, Karen, and Aaron Dimock. "Meetings: Discursive Sites for Building and Fragmenting Community." In *Communication Yearbook*, edited by Pamela J. Kalbfleisch, 28 (2004): 127–166. Mahwah, NJ: Lawrence Erlbaum Associates.

Tracy, Sarah J., Karen K. Myers, and Clifton W. Scott. "Cracking Jokes and Crafting Selves: Sensemaking and Identity Management among Human Service Workers." *Communication Monographs* 73, no. 3 (2006): 283–308. doi:10.1080/03637750600889500.

Whittaker, Steve, and Brid O'Conaill. "The Role of Vision in Face-to-Face and Mediated Communication." In *Video-Mediated Communication*, edited by Kathleen E. Finn, Abigail J. Sellen, and Sylvia B. Wilbur, 23–49. Mahwah, NJ: Lawrence Erlbaum Associates, 1997.

Yamada, Haru. "Topic Management and Turn Distribution in Business Meetings: American Versus Japanese Strategies." *Text—Interdisciplinary Journal for the Study of Discourse* 10, no. 3 (1990). doi:10.1515/text.1.1990.10.3.271.

Yamada, Haru. *American and Japanese Business Discourse: A Comparison of Interactional Styles.* Norwood, NJ: Ablex, 1992.

Yamada, Haru. "Organisation in American and Japanese Meetings: Task versus Relationship." In *The Language of Business: An International Perspective*, edited by Francesca Bargiela-Chiappini and Sandra J. Harris, 117–135. Edinburgh, UK: Edinburgh University Press, 1997.

FOUR

Delving Deeper
into Online Peer Feedback

Implications for Product Design

Maaike Bouwmeester

Increasingly, teacher preparation programs are using blended learning approaches to extend and deepen the relationships and communities of practice that form among student teachers (STs). Given the limited face-to-face time afforded by the classroom environment, growing numbers of STs are trying to expand their learning environments into the realm of online interactions (Lin 2008). Often these environments build on and complement face-to-face interactions between peers as they discuss, commiserate and problem-solve around the challenging and novel situations they encounter in classroom settings. Online tools and asynchronous peer interactions provide important spaces for STs to further develop ideas, raise concerns, discuss the challenges they face, and strengthen a sense of community among peers.

This was true for a graduate Science Education program at an East Coast University (ECU) in the United States where, along with another faculty member, I co-designed and studied an intervention that provided online opportunities for peer STs to share and respond to reflections and questions that came up for them in their student teaching experiences. I served as the principal researcher of this online learning activity, which was the basis for my dissertation research, completed in 2011. Using a case study approach that included extensive interviewing, observation, and an eight-month immersion in the culture and activities of this teacher education program, I evaluated the efficacy of the intervention and also explored theoretical questions that examined the role of reflection in helping students make connections between concepts and strategies learned in coursework with their field experiences (for example, student teaching and observations).

In addition, as one of the primary product developers of the Streamline software[1] being used in this activity (described in more detail herein), I was interested in learning how the technology supported, altered, or constrained the discourse and relationships between peers. Since teacher education students represent the vast majority of users of Streamline's online learning and assessment products, the product development team at Streamline hoped to use findings from the study as potential inputs into prioritizing and developing product modifications.

The purpose of this chapter is not to recapitulate the findings from the original study. Rather, my intention is to re-examine a subset of the original research questions and data from a cultural discourse perspective in order to better understand how conventions, shared practices and cultural norms of "relating" within this particular group of STs shaped their experiences of the activity. By revisiting the study from this perspective, this chapter explores some potential new interpretations of users' experiences that emerged when local context and cultural norms were taken into consideration. My hope is that this chapter will contribute to the application of cultural discourse analysis methods by researchers and product designers, who may benefit from taking a more holistic approach to interpreting users' experiences with their web-based applications.

PRODUCT DESIGN

In her book, *Speaking Relationally: Culture, Communication, and Interpersonal Connection* (1998), Kristine Fitch writes "cultural values and norms have been offered with increasing frequency as explanations for people's actions, from greeting sequences to arranged marriage, and for the way people interpret the actions of others" (2). While this quote may represent something close to conventional wisdom within the ethnography community, it has been my experience that many web-based product and interaction designers often fail to consider cultural values and norms when interpreting users' experiences with their products. For example, even commonly used web-based forms often fail to adhere to simple "netiquette" around the use of informal and formal pronouns (for example, the use of tú and usted), a distinction that connotes respect and politeness, especially in Central and South America. Norms around language are but one example of how attention to culture can influence how users experience an online environment.

Janet Murray defines interaction design as including "many aspects of the system that have to be the subject of coordinated design decisions, including social and cultural elements as well as technical and visual components" (Murray 2012, 11). While most product and interaction designers do identify typical group attributes of their users (often developing archetypal "personas," as described in the introduction to this book), they often don't examine the underlying cultural norms and values that may have equal import to the design of interactions and user interfaces.

I recognized this while working as VP of Product Development at a growing and dynamic product development organization. In the fourteen years I worked there, I hired and worked closely with multiple product managers, business analysts, visual and interaction designers (labeled "product designers" here on forward to simplify the often overlapping functions and titles). Over time, we experimented with and adopted various methods to aid us in developing effective and efficient design practices that included discovery, customer validation, prioritization, and design processes. Like many companies, we ended up adopting a rapid and iterative approach called SCRUM which is associated with Agile development methodology (PCW 2014).

As a result of early customer feedback and rapid development cycles associated with an Agile process, product designers can usually quickly identify and fix users' specific frustration points (also referred to as "pain points" in the industry). They often do this by modifying or adding visual design elements, interactivity, or features. However, product designers often look at individuals' circumstances or a group's overall concerns without looking at deeper cultural inferences that may also play a role in a user's experience. This book argues that understanding this social and cultural context can provide additional layers of meaning and insight which in turn can shed light on previously undiscovered but needed modifications or even entirely new products or features.

PEER RELATIONSHIPS IN TEACHER PRACTICE

Among the myriad factors that contribute to teacher learning, pre-service teachers often find field experiences to be most important (Wilson, Floden, and Ferrini-Mundy 2001). Field experiences include observing classroom teaching events and actually teaching in a K–12 classroom with oversight from the teacher of record for that class (the "cooperating teacher"). STs are mentored and guided by program faculty and cooperating teachers who help STs make sense of what Schön (1992) describes as "uncertainty, uniqueness, and conflict. The non-routine situations are at least partly indeterminate and must somehow be made coherent" (157).

Given the time-intensive nature of one-to-one mentorship by the faculty and teachers, STs in teacher preparation programs often form close relationships with peers who are going through similar experiences—often sharing tips, observations, frustrations, and accomplishments. Korthagen

(2001) writes: "By sharing experiences, [student] teachers are stimulated to structure these experiences and, by comparing their own analysis of practice with those of others, they may discover other possible ways of framing their experiences" (149). These peer bonds and communities often form inside and out of class, but can be further nurtured via online activities that promote increased sharing and interactions between peers and instructors (Lock 2006).

STREAMLINE'S E-PORTFOLIO
AND COMMUNICATION CAPABILITIES

In a review of the literature on blended learning in teacher prepara-tion programs, Keengwe and Kang (2012) found that faculty in teacher preparation programs are increasingly using online communication tools to promote discourse activities among peers. Most typically use simple online discussion tools such as Blackboard, Ning, and Facebook, but some are using more advanced online sharing experiences that include other learning activities as well. Khine and Lourdusamy (2003), for example, re-ported on uses of online forums to help STs brainstorm around problems they encountered in their teaching practice. Lin (2008) stated that students often found online discussion boards to be more useful than in-class discussions because students could take the time to compose a response.

Another example is teacher preparation programs that use e-Portfolios to help STs document and share examples (or "artifacts") of their learning process—including artifacts such as lesson plans, case studies, videos of student teaching, and written reflections. Based on an internal marketing survey conducted by Streamline in 2014, 60 percent of teacher preparation programs were using e-Portfolio tools in some capacity. At the time of data collection, Streamline served over half of this market.[2]

Streamline is one of several popular e-Portfolio tools on the market and in wide use by teacher preparation programs in the United States. Streamline was selected by the ECU Science Education program for use in this study after I suggested its use based on my familiarity with the tools, but also because the tools could facilitate the kind of peer interactions and communication capabilities mentioned above. e-Portfolios—much like a personal website for presenting work—are created by students and include a compilation of artifacts, often accompanied by reflections or an-notations on what the artifact is, the context for its development and what was learned in the process of creating it. Teacher preparation programs use e-Portfolios for multiple purposes: at their most basic, for showcasing ST work for purposes of employment, certification, or credentialing, but equally, if not more importantly, for learning and assessment.

Zeichner (2010) writes, "given the increased emphasis on preparing and developing teachers to be reflective and analytic about their practice, teacher educators and staff developers have increasingly used teaching portfolios as one of a number of vehicles to stimulate greater reflection and analysis by teachers" (91). As a learning tool, portfolio artifacts can be shared as the basis for discussion and formative feedback with peers and instructors (Mansvelder-Longayroux, Beijaard, and Verloop 2007). Zubi-zarretta (2009) writes that the portfolio may "improve student learning by providing a structure for students to reflect systematically over time on their learning process and develop the aptitudes, skills and habits that come from reflection" (15).

As demand for e-Portfolios has increased, commercial and open source e-Portfolio software has proliferated along with many new communication features that enable students to share and respond to e-Portfolio work with others. Because of these additional features, creating a social learning environment around portfolio work is much easier now than in earlier manifestations of the portfolio. STs, peers, and instructors easily exchange feedback by adding comments or evaluations or by engaging in discussions focused around their work.

Though the use of peer review to respond to reflective work is a less common use of the Streamline tools, this use of the tools was of particular interest to Streamline because of the emphasis that many teacher preparation programs place on fostering reflective practice among teachers. Providing peers with opportunities to share and respond to reflective writing often has benefits for both the reviewer and reviewee. For example, Topping (1998), in a review of the literature, examined thirty-one studies on peer feedback and found that most showed that peer activities improved the learning for both the reviewer and the reviewee. The studies reviewed in this literature review also indicated that peers were able to give better feedback when they had prompts or guidelines on what feedback to give.

In addition to commenting features, Streamline includes the ability to create and attach rubrics that can be used in a variety of ways. Rubrics articulate the expectations for an assignment by listing the criteria of things important to consider, as well as describing levels of quality of each of those criteria (much like a Likert scale). While rubrics are often used for grading, they can also be used for learning purposes rather than assessment. In this study, rubrics were designed collaboratively with students to be included in the online activity as "reflection prompts." A reflection prompt is a scaffold or reminder that aids students in thinking of things that *could* be included in their reflective writing. The use of prompts to focus student thinking is a common scaffolding technique. Without this structure, educators have often reported that students have difficulty reflecting on their own, though as students become more familiar with

	Descriptive	Elaborative	Transformative	Score/Level
Reflections on course materials and discussions	Descriptions and observations are noted, but no further analysis or insight provided	• Elaborates and synthesizes course materials and discussions • Shares personal reactions (agreements/ disagreements) • Develops new insights	• Draws on multiple perspectives. • Recognizes broader context • Asks hard questions that reframe perspective and challenge personally held assumptions about course materials and discussions.	
Connections between course materials and discussions AND experiences in the field	Makes connections, but no further analysis or insight provided	• Elaborates on connections • Shares personal reactions (agreements/ disagreements) • Develops new insights based on these connections	• Draws on multiple perspectives in making connections • Recognizes broader context in which connections can be made. • Asks hard questions that challenge personally held assumptions.	

Figure 4.1. Reflection Rubric. Data collection and figure creation by Maaike Bouwmeester.

reflective writing, they often prefer to do it without prompts (Moon 2004). In the e-Portfolio tools in Streamline, we included a rubric (see Figure 4.1) to embed prompts in key areas where reflective writing and peer responses were required. These prompts guided students to focus on key criteria as they wrote up their reflections and similarly as peers responded to those writings.

METHODS

The data collection for the original study took place in 2011 during the spring semester of a Methods course in the Science Education masters program at ECU. This case study included six students and their instructor in the Methods course mentioned previously. The course was designed to promote reflection and knowledge acquisition related to topics relevant to student teachers in their teaching practice including content knowledge, pedagogical knowledge and other skills such as classroom management skills.

Data originated from in-depth interviews, observations, and field notes of classroom sessions, as well as through weekly reflective writings and peer responses that students added to their e-Portfolio and shared with each other and the instructor using an online e-Portfolio system. For more details of methods and protocols I used in the original study, please refer to my dissertation (Bouwmeester 2011).

For the purpose of this chapter, I revisited my interview transcripts and field notes and analyzed the data using the "interpretive mode of inquiry" described by Donal Carbaugh (2007), who writes:

> The interpretive mode responds to the question: what is the significance and importance of that phenomenon to participants? Or, in other words, what meanings are active in this communication practice? The task here is to provide an interpretive account of the practice, identifying the premises of belief and value that are active when one does such a thing. What needs to be

presumed, or understood, in order for this kind of communication practice to be intelligible here? (172)

In reinterpreting my data from this perspective, I decided to look more closely at the types of peer experiences STs were having prior to participating in the activity, with the goal of understanding how existing cultural norms and values were supported, altered, or even possibly undermined by the activity of interacting online as described in this study. Prior to the official data collection phase described above, I had observed a number of classes in the program, interviewed several instructors and sat in on an earlier version of the methods course in order to become familiar with ST experiences, curriculum, and other dynamics in the program. Additionally, one of my research questions had focused on peer feedback and I had a plethora of data to go back to which I excerpt in subsequent sections.

It is worth noting that I am using data collected from *written* discourse between peers as well as self-reported data via the interviews I conducted. When this happens there is the possibility that what peers said contradicts what I observed in their written online communications. On this, Carbaugh (2007) writes:

> There is a difference between analyzing meanings-in-practice, as part of on-going social interaction, and analyzing participants' reports about that practice. Each involves different orders of data, the former being an enactment of the practice, the latter a report about it. While these can complement each other, they can also diverge. A participant can deny some interpretations of a practice, even though those interpretations are robust in social interaction. If this is the case, that is, a discursive practice has competing meanings, some amplified while others muted, the analyst wants to know this! The range of active meanings in and about the practice is thus the target of the interpretive analysis. (174)

In sum, in writing this chapter I used a particular mode of inquiry described as the "interpretive mode" by Carbaugh (2007), to examine the ways in which ST self-reported answers and written discourse with peers helps us understand how they understood themselves, the value of peer relationships and how those relationships were influenced by the technology that mediated those relationships.

RELATING AS PEERS ONLINE AND OFF

Interviewer: How do your interactions with peers right now differ from your previous academic experiences?

Interviewee: So at ECU everyone in my class is always sharing what they have to say and they're always very opinionated. They always have their own opinion and I liked that. It opened me up to be able to say what I have to say.

Interviewer: Can you remember or share any examples of, like, times when your peers have been in some way influential on what you're doing in student teaching?

Interviewee: Yeah. So, in all our classes we always talk about like the highs and lows of the week or like something that we've experienced through student teaching that helped us or some kind of strategies that we used in our classroom that were effective. I can't remember specific examples, but they have been useful to me in terms of like—I can't remember if I used one of the other strategies that other classmates have used, but it helps me think about other strategies and other methods that I can use in the future. So that's useful. (Quote from Amanda [interview #1—before participating in the online peer exchange])

Amanda's positive view on her peers was a shared sentiment among all the STs I interviewed, and was corroborated by my field notes on face-to-face class exchanges I observed. Peer sharing, as it occurred in classes and expressed through spontaneous conversation outside of class, seemed to be a highly valued practice among STs in this program. Peer relationships allowed STs to exchange opinions and strategies and discharge some of the stress and uncertainty that STs experienced in their student teaching experiences.

The design of the online peer exchange activity, as previously discussed, was largely based on the assumption that peers would welcome and benefit from opportunities for extended and semi-structured interactions outside of class time. The analysis, however, did not support that assumption. While some students did value the activity and participated enthusiastically, others experienced it as a waste of their time.

In examining peers' written feedback and interview data, I found that students could be placed into one of two archetypes or "personas": the "benevolent mentor" and the "disengaged skeptic." The benevolent mentor provided feedback that was empathetic, detailed, and productive. These students typically valued communication and had reflective dispositions (that is, "I need to practice good feedback so that I get better at it"). The "disengaged skeptic," by contrast, tended to feel that peers (including themselves) weren't experienced enough to provide useful feedback and that only a more experienced mentor could provide the kind of feedback they valued. They tended to place less emphasis on the value of social construction of knowledge and tended to give minimal feedback and place less value in the feedback they received. Excerpts and data analysis provide examples of these two sentiments.

Vanessa

Vanessa, one of the more prolific writers among the group, gave thoughtful and substantive feedback to her partner, Paul. In responding to Paul, Vanessa assumed the role of a benevolent mentor or teacher: providing positive affirmation of the opinions he expressed but also prodding him to extend his thinking.

> I can see that you are starting to form ideas about what it will mean to implement labs in your own future science classroom. When you're working one-on-one with students, what scaffolding techniques or other strategies do you use to support your students? How do you think these one-on-one techniques could transfer over to a 1 to 30 ratio? Maybe it's not possible; what other strategies could work?

Vanessa tried to get Paul to probe more deeply into the topics he was writing about. Her answers to interview questions were consistent with my observation (for example, "I try to ask more questions. Like, 'You said this, but what informs that?'"). In some cases, she provided examples from her own experiences to help him develop deeper insights. For example, in responding to Paul's comments about class size, Vanessa included some ideas by the founder of the school at which she was student teaching. She wrote:

> Overcrowding/large class sizes are a challenge that teachers have to face. Progressive educational leader, Theodore Sizer (founder of the Coalition of Essential Schools, ex: Vanguard, ICE, School of the Future, etc.) wrote that a teacher shouldn't have a student load larger than 80 in order to know all his/her students, but schools cannot fulfill this ideal because they are at odds with budget cuts. So, I liked that you reflected about the importance of developing a relationship with your students.

Vanessa brought a new angle to a topic that Paul was writing about; she pointed out that the ideal of small class size is often at odds with the realities of school budgets and funding.

Amanda

Amanda, by contrast, gave relatively sparse feedback compared to the more extensive feedback from Vanessa. Amanda stated that she did not particularly like the online peer exchange for several reasons. One was that she guessed that peers were unlikely to spend quality time giving feedback because they would want to get it over with quickly. She also did not think peers had enough experience to give quality feedback. In contrast, she found the instructor, Jonathan's, feedback to be very valu-able. She said:

> There is a significant difference [between peer and Jonathan] because Jona-than has experience, and he has seen all this before maybe ten times over. And he's been in the classroom, and he knows each school's personality. So his feedback and perspective is much different. It's very big picture, up here I can see everything kind of perspective. Whereas my peer is like well, that may be happening at your school, in your classes, but I personally have never seen it. So, I don't know how to advise you on it or how you would approach a particular situation because I haven't seen it.

Looking at this data from a purely individualistic perspective as I did in my initial analysis, I could see differences in personality traits of individ-ual students that were similar to Korthagen's (2001) assertion that teacher education students often have very different learning orientations.

> Internally oriented students want to learn by reflecting on their experiences, and externally oriented student teachers want instructions and guidelines from the teacher educator. If students of the latter type are forced to learn in a manner that is not theirs, then they may feel they are putting a lot of time into something they do not consider useful. (228)

However, I was perplexed that Amanda, who expressed that she valued spontaneous informal face-to-face exchanges that came up in classroom or informal contexts ("it helps me think about other strategies and other methods that I can use in the future. So that's useful"), found it of so little use in the online peer exchange. In the written online peer exchange, Amanda preferred to receive comments and feedback from a highly knowledgeable and experienced reviewer. She felt that Jonathan's experience was essential in providing her with useful feedback because he "has seen all this ten times over."

It seems that some students experienced these online interactions very differently from others. What cultural norms, values, and beliefs might have contributed to this disparity?

With this question in mind I went back to my data, going beyond the individual level of analysis I had done in the original study to look at how the norms around social interaction may have been at odds with the design of the activity and/or the Streamline peer review features as expressed through the interface and interaction design. There were two prominent features that stood out: 1) in expounding on their discomfort with the online exchange, some STs expressed concern that the activity felt like an assessment activity; and 2) students felt constrained by the 1:1 pairing we had set up (versus other formats they had experienced that were more open). Both of these observations are explored in more detail in the next two sections.

PEER SUPPORT OR PEER ASSESSMENT?

Janet Murray writes that "in order to make truly intuitive interfaces, designers must be hyperaware of the conventions by which we make sense of the world—conventions that govern our navigation of space, or use of tools, and our engagement with media" (2012, 11). This appeared to be especially relevant for student teachers using the Streamline platform. As mentioned above, the online peer exchange was mediated by the use of a rubric which was used as a reflective prompt to focus student thinking, help them think of things to write about and similarly, to prompt peers on how they might respond to their partners' writings.

The students' activities in the peer online exchange did not count towards the grade for the class. The instructor and I reinforced this point to avoid having STs perceive that the rubric was a grading rather than a learning tool. Students were also told that they could ignore the rubric and should use it only if it was helpful to them. This instruction was reiterated several times throughout the semester.

However, since rubrics are more *typically* used for grading students than as an informal prompt to guide communication, the designers of the Streamline rubric capabilities had used conventions (wording, features, etc.) associated with assessment. Words like evaluate, performance level as well as actual numeric scores that denote the different "levels" of performance (usually from poor to excellent) are conventions within the user interface.

One of the interesting findings that emerged from the interviews was that while students understood that their reflections, feedback, and use of the rubric were not being graded, peer reviewers couldn't overcome the feeling that they were in fact grading their peers, something they were deeply uncomfortable with given the importance of their peer relationships in the program. For example, one student reported: "Even if I don't care what grade I would have gotten but just by like by virtue of saying one, two, three, and we all have been in school for a long time, whether we like it or not grades kind of come into our lap."

Another student said simply: "no matter what you wish to think, you do see it as a grade." Though these two students knew they weren't being asked to grade their peers, on an emotional level (reinforced by many years of being graded), the representation of scores and evaluation conventions used on the interface gave them the *feeling* that they were grading their peers. They were also sensitive to how their feedback might be interpreted by their peers if it seemed that they were grading versus just offering feedback.

My initial reaction to these findings was to take this user feedback at face value and recommend that the Streamline product design team "fix" the particular language that was causing the presumed *problem*. Indeed many product designers take user frustration and "pain points" at face value, making changes that seem to address the difficulty experienced by users while failing to recognize a more systemic misalignment between users' experiences and the design of the system.

In this case, I started considering that STs arrive in teacher preparation programs with years of experience observing their own teachers and being indoctrinated into a culture that values grades. Dan Lortie (1975) labeled this effect the "apprenticeship of observation" and the concept is useful in helping to understand how STs' entering beliefs shape their ideas and practices. Prior experiences as students and exposure to teacher stereotypes may constitute beliefs that exert a powerful influence on what pre-service teachers observe and experience in their education programs, as was likely the case here. Some of the STs were simply unable to shake the feeling that they were "grading" their peers and that this undermined the mutually respectful nature of those relationships.

PAIRED VS. OPEN REVIEW

Some STs reported that they preferred a more open online format for peer exchange, such as those typically found in discussion boards and social media sites. A review of my interview transcripts and field notes helped me better understand how culture and context may have played a role in shaping their communication preferences.

In semesters prior to the study, STs had used an online discussion board as a supplementary communication channel in a number of their courses. This forum allowed them to add comments, and read and respond to others' comments. Comments were publicly visible to all while the peer partnering introduced in this activity limited the flow of conversation to two people. At the same time, the partnering format also (intentionally) facilitated a more structured exchange of ideas as a result. Indeed, the whole feedback process in the paired activity, including the weekly submission of writing and feedback guided by the reflective prompts, was a much more structured process. In contrasting these two experiences, some students found value in a more open exchange, while others preferred the more structured partnered pairing.

Lilly, for example, wrote that the peer partnering felt "forced." Another student reported that "when I used Blackboard [discussion board], when

I gave feedback I usually just agreed with the person. With Streamline, I had to think about how to make my feedback to her meaningful so that she could learn from it." Other students said they saw the value of both but craved the openness of the discussion board where they could communicate with multiple peers rather than just one.

On the one hand this was not a surprising finding. STs—and, indeed, most teacher educators—value the participation in what Wenger (n.d.) terms communities of practice: "groups of people who share a concern or a passion for something they do and learn how to do it better as they interact regularly" (1). These types of communities are considered by almost all teacher educators as essential for learning the craft of teaching and most teachers participate in some form of community for the purposes of learning with pleasure. Wenger points out the definition of communities of practice "allows for, but does not assume, intentionality, learning can be the reason the community comes together or an incidental outcome of member's [sic] interactions" (1).

In this sense, an open forum like a discussion board might serve to nurture and support communities of practice while for others the intentionality of a structured process might be a better method to achieve this. The STs' reactions were indicative not only of their individual preferences but also underlying beliefs and norms around community and learning.

In designing products and features that facilitate relationships and communication, product designers are wise to think about the ways in which features enable either open or structured interaction. For example: does the system enable only one-to-one transactions (for example, instructor-student, student-student) or does it also allow for one-to-many or many-many transactions (for example, social networks, discussion boards)? This same consideration applies to conferencing software discussed in the previous chapter; seemingly small changes to product features and preferences can result in significant differences in how people can interact and form communities.

CONCLUSION

In cultural discourse analysis, researchers examine communication events and situations to better understand participants' conceptions of identity, action sequence, emotion, sociality, and location (Scollo 2011). In this study, the researcher focused on the ways in which an online peer exchange revealed STs' underlying values, beliefs, and preferences.

By taking such an approach, I started to appreciate that the design of digital products is indeed what Murray (2012) describes as very much like a "cultural practice like writing a book or making a film" (1). Even when designers cannot initially envision the multiple uses of their products, they benefit from user feedback not simply to suggest modifications to specific features and designs but also to reveal important considerations about how context of use and cultural norms are being mediated with and through the use of the technology. As Murray (2012) explains, "A digital telephone can be understood as a neutral piece of business equipment, a stylish status symbol, an intrusive disruptor of family or a lifeline. The humanist designer aims to see as much of this larger web of meaning as possible in order to understand the context and connotations of particular design choices." (1)

In the end, this particular online peer activity did not seem to draw peers or the larger ST community closer to one another, and for some STs it even introduced some tension, especially around the idea of grading one's peers. Luckily this was a single activity (out of many peer sharing opportunities they would encounter in the program) so it is likely that these STs weathered the discomfort without lasting strains to their community and peer bonds.

In this chapter, I hope to have provided a useful example that illustrates how an understanding of cultural context and the study of discourse can provide unique insights that can help inform the product design process. By understanding how these learners saw themselves in relation to the learning process, as well as how they expected to relate to one another, it was possible to identify not only product enhancements, but also to develop a deeper understanding of users' views on the activities and features themselves. Taken together, these findings point towards an opportunity for a more nuanced, culturally grounded approach to product development than the typical feature-driven approach favored by many commercial software companies. This interplay between cultural discourse analysis and commercial product development could provide fertile ground for further exploration.

NOTES

1. Streamline is a pseudonym. All identifying references have been removed.
2. Internal Streamline marketing research, conducted in 2014.

REFERENCES

Bouwmeester, Maaike. "Examining the Effects of Reflective Rubrics in the E-Portfolio Peer Review Process on Pre-service Teachers' Ability to Integrate

Academic Coursework and Field Experiences." PhD dissertation, New York University, 2011. ProQuest (UMI 3478271).

Carbaugh, Donal. "Cultural Discourse Analysis: Communication Practices and Intercultural Encounters." *Journal of Intercultural Communication Research* 36, no. 3 (2007): 167–182. doi:10.1080/17475750701737090.

Fitch, Kristine. *Speaking Relationally: Culture, Communication, and Interpersonal Connection.* New York: The Guilford Press, 1998.

Keengwe, Jared, and Jung-Jin Kang. "Blended Learning in Teacher Preparation Programs: A Literature Review." *International Journal of Information and Communication Technology Education* 8 (2012).

Khine, Myint Swe, and Atputhasamy Lourdusamy. "Blended Learning Approach in Teacher Education: Combining Face-to-Face Instruction, Multimedia Viewing and Online Discussion." *British Journal of Educational Technology* 34 (2003): 671–675.

Korthagen, Fred A. *Linking Practice and Theory.* New Jersey: Lawrence Erlbaum Associates, 2001.

Lin, Hong. "Blending Online Components into Traditional Instruction in Pre-Service Teacher Education: The Good, the Bad, and the Ugly." *International Journal for the Scholarship of Teaching and Learning* 2, no. 1 (2008): 1–14.

Lock, Jennifer V. "A New Image: Online Communities to Facilitate Teacher Professional Development." *Journal of Technology and Teacher Education* 14 (2006): 663–678.

Lortie, Dan. *Schoolteacher: A Sociological Study.* Chicago: University of Chicago Press, 1975.

Mansvelder-Longayroux, Désirée Danièle, Douwe Beijaard, and Nico Verloop. "The Portfolio as a Tool for Stimulating Reflection of Student Teachers." *Teaching and Teacher Education*, 23 (2007): 47–62.

Moon, Jennifer A. *A Handbook of Reflective and Experiential Learning.* New York: Routledge Falmer, 2004.

Murray, Janet H. *Inventing the Medium: Principles of Interaction Design as a Cultural Practice.* Cambridge, MA: The MIT Press, 2012.

PCW. "Adopting an Agile Methodology Requirements—Gathering and Delivery" (2014). Accessed February 6, 2015. https://www.pwc.com/en_US/us/insurance/publications/assets/pwc-adopting-agile-methodology.pdf, retrieved 2/6/2015.

Schön, Donald A. *The Reflective Practitioner.* San Francisco: Jossey-Bass, 1992.

Scollo, Michelle. "Cultural Approaches to Discourse Analysis: A Theoretical and Methodological Conversation with Special Focus on Donal Carbaugh's Cultural Discourse Theory." *Journal of Multicultural Discourses* 6, no. 1 (2011): 1–32.

Topping, Keith. "Peer Assessment Between Students in Colleges and Universities." *Review of Educational Research* 68, no. 3 (1998): 249–276.

Wenger, Etienne. "Brief Introduction to Communities of Practice" (n.d.). Accessed February 9, 2015. http://wenger-trayner.com/introduction-to-communities-of-practice.

Wilson, Suzanne, Robert E. Floden, and Joan Ferrini-Mundy. *Teacher Preparation Research: Current Knowledge, Gaps, and Recommendations.* Seattle, WA: Center for the Study of Teaching and Policy, 2001.

Zeichner, Ken. "Rethinking the Connections between Campus Courses and Field Experiences in College- and University-Based Teacher Education." *Journal of Teacher Education* 61 (2010): 89–99.

Zubizarretta, John. *The Learning Portfolio: Reflective Practice for Improving Student Learning*. San Francisco: Jossey-Bass, 2009.

FIVE

The Code of WeChat

Chinese Students' Cell Phone Social Media Practices
Todd L. Sandel and Jenny Bei Ju[1, 2]

Many city governments across greater China post public service announcements that are intended to guide behavior in public places. These include such signs as fines for smoking or littering, warnings not to spit or chew betel nut. When riding public transportation, such as the subway or bus, there are reminders to offer seats to the elderly, disabled, women who are pregnant or carrying young children, and directions where to stand when boarding and alighting. Because the flow of traffic is different across the region—vehicles travel on the left in Macao and Hong Kong, but on the right in mainland China and Taiwan—at some busy intersections pedestrians are reminded which direction to look for oncoming traffic. Such signs are both a reflection and commentary on the challenges of contemporary urban living, places where millions of people are in close contact with each other. These signs reflect the disruptive processes of rapid urbanization on people's lives and behaviors. When people move from rural to urban environments, an assumption is that people must be "taught" that behaviors acceptable in one context are unacceptable in a new one. Evidence for this is seen in both government-sponsored educational campaigns across greater China, and in local folk understandings of the faults of "outsiders" as people who "lack education"; this we see in how local Macao people perceive the habits and actions of mainland Chinese tourists (see Guan and Sandel 2015).

While the examples of public service signs may be aimed at "educating" people who bring rural habits to urban areas, recent years have witnessed the rise of an urban-based habit that disrupts the flow of people—and in some places warrants a public service sign. In Chinese, these people are known by the phrase, *di tou zu* 低頭族, or literally "head lowered tribe."

This phrase describes the actions of people who lower their heads as they look at and manipulate the screens of their cell phones. In Hong Kong travelers are warned through both public announcements and printed signs not to look at their cell phones while entering or alighting from escalators. In Taiwan, at a recently opened subway station, a directional marker for how to transfer to other subway lines was painted on the floor in large, bright colors. The headline on a news report notes that it is so big and bright that even "the head lowered tribe can see it" (Lee 2014).

Social behavior in public places is changing in China's urban areas, and such change is motivated in part by the increasing use of cell phones. Such devices impact and change ways that people walk, talk, communicate, and interact with others on a daily basis in multiple ways. In this chapter we focus our attention on one aspect of these changes, the rules, or communicative codes, that university students in China have developed when using WeChat (known as *Weixin* 微信 in Chinese), one of the most popular social media platforms in China. Launched in 2011 by the Tencent Company, WeChat has quickly grown in popularity (Hou 2014); by the middle of 2014 it had more than 400 million active users (Hong 2014). While the application is similar to Facebook, as users may exchange instant messages, and post and comment on pictures, it differs in a number of ways. Our interest in studying WeChat is not only in describing how people use the application, but also discussing an emergent set of rules, or social codes that users have developed for appropriate ways to communicate.

Theoretically we draw upon work in the study of communicative codes, beginning with that of Dell Hymes (Hymes 1974), and subsequently developed as a theory of speech codes by Gerry Philipsen and colleagues (Philipsen, Coutu, and Covarrubias 2005). Hymes claimed that distinct groups of people, whom he referred to as constituting "speech communities," shared an understanding of what counts as communication, and the rules for its use (Hymes 1974). For example, Hymes explains that the "Wishram Chinook of the Columbia River in what is now the state of Washington . . . have considered infants' vocalizations to manifest a special language. . . . [T]his language was interpretable only by men having certain guardian spirits" (31). That is, the Wishram not only possessed a unique understanding of what counts as communication—infants' vocalizations—they also had a rule for defining who and under what conditions such communication could be understood—men with certain guardian spirits.

Subsequent work by Gerry Philipsen and colleagues draws upon Hymes's insights to further our understanding. In an ethnographic study of a community called "Teamsterville" he identified a "code of honor" that was at play in everyday interactions. For example, talk was highly gendered with distinct places for men's and women's talk, and intermediaries were depended upon for talk with outsiders and/or people with higher status (for example, a priest would speak to God on someone's behalf) (Philipsen 1992). Subsequent work demonstrates that interactants develop informal rules for appropriate behavior in social situations (Milburn 2004), and that these rules are multiple and vary across contexts (Homsey and Sandel 2012). This has been developed into a theory of speech codes with six propositions that provide a map for how to identify, analyze, and evaluate speech codes (Philipsen, Coutu, and Covarrubias 2005). While this line of work has provided us with many insights about the nature of communication in face-to-face or "offline" communities, less work has been done to understand the communicative or speech codes of online and/or virtual communities. The present study is intended to fill a gap in our understanding about one emerging set of cultural codes for behavior among one group of participants—Chinese university students—who use WeChat, one new social platform.

The remainder of this introduction proceeds as follows. We begin by reviewing relevant studies of codes of communication in online or virtual communities, demonstrating how they may develop and how they may differ from codes found in traditional face-to-face communities. This is followed by an overview of a number of studies that describe WeChat as a technology and communicative resource. While none examine WeChat as a communicative code, these studies do help us understand how and why users may communicate via WeChat.

ONLINE COMMUNITIES

In earlier work on computer-mediated-communication, Walther (1992) claimed that relational messages can be textually conveyed online. That is, people use resources at hand—verbal and nonverbal—to communicate relational closeness or distance. For instance one verbal resource for communicating relationally can be seen in "flaming": Language that is in the form of "insults, swearing, and hostile, intense language" (56). Relational affiliation may also be conveyed nonverbally through the use of "paralanguage . . . intentional misspelling, lexical surrogates for vocal segregates . . . capitalization" (79). Later work has found that users develop new vocabularies and language to communicate messages online. For instance, in a study of IM (instant messaging) conversations Baron (2004) found users developed an assortment of abbreviations and acronyms (for example, cya = see you, brb = be right back, lol = laughing out loud) and emoticons (for example, :-) = smiley, :-P = sticking out tongue) that they blended with written language.

The above show that online communication is not necessarily inferior to face-to-face communication. Instead, the two are different, and operate under different conditions, as claimed by Baron (2004): "CMC has its own usage conditions, and therefore, each needs to be analyzed in its own right. These usage conditions may, in turn, influence the character of language produced in that medium" (398). For example, the dimensions of time and space are perceived differently in online environments. In asynchronous communication (for example, email), a person may show close affiliation by responding quickly. Or, in an interaction whereby one person is offended by an online message, the other may not know this for a period of time, absent the facial and nonverbal cues that would be available face-to-face. In addition, with the development of social media platforms, such as Facebook, messages are posted that may be read by multiple users, and may be retrievable over an extended period. This means that such users may have to consider potential reactions from a wider audience than would be the case in traditional, face-to-face communication.

Now consider what can be learned from studies of the communicative codes found in online communities. One of the earliest studies is described by Fitch (1999). In the 1990s an academic community of scholars developed a Listserv—a text based online discussion group—and began to exchange messages. Soon after its inception one person "posted a message that both flamed and inflamed the newly formed . . . community" (42). This participant related a story of a student who attended a seminar with someone she called her "friend," who was later revealed to be her husband. Others objected, saying that the term friend "excluded that he was her husband." A few hours later, another participant described terms that she used to describe close relational partners (for example, significant other, partner, lover) and then concluded with the following: "When was the last time you read an OBIT which said fuck-pillow and so on?" The phrase "fuck-pillow" sparked a lively and heated online discussion among Listserv members. Fitch claims that they were struggling to "establish a common code," or a way to frame the bounds of proper communication as they discussed both "substantive issues" regarding terms for describing relational partners, and "metatalk" about such terms (44).

In their discussion of the language of social media, Seargeant and Tagg observe that the concept of a "speech community" began with the assumption that it is comprised of mostly like-minded persons who interact in a physically embodied context on an everyday basis, much like that observed by Philipsen in Teamsterville (Seargeant and Tagg 2014). Yet with the emergence of online and virtual communities, such an assumption no longer applies. In the era of globalization new factors impact the development of communities, namely: "mobility, the range of different affiliations people have, and the dynamic nature of language use in general" (11). They explain that networked connections may be more "flexible, shifting, and interactively constructed" than offline ones (11). Time and space constraints do not matter as much as they do in physical contexts. Furthermore, social media platforms such as Facebook allow both one-to-one and one-to-many communication messages. That is, a person can choose to post a picture and comment on a friend's wall, thus keeping the conversation private. Or, the same posting can be made public for all friends to see. Thus, online communication allows users to more easily switch audiences than is possible in traditional face-to-face interaction.

WECHAT

We now turn our attention to discuss one of China's most popular social media platforms, WeChat. This is preceded by a brief narrative of the development of social media in China and the issue of censorship as it helps us better understand the culture of China's online communication.

According to Yang (2009) China "achieved full-function connectivity to the Internet in 1994" (2) and Chinese people quickly and readily embraced the Internet as a forum that could provide a wider range of self- and community-wide expression. A distinctly Chinese online culture emerged, one that demonstrated a "creative culture full of humor, play, and irreverence" (2). Unlike the offline, face-to-face world where interactants must carefully be attuned to the social order and attempt to maintain a surface level harmony (Chang 2010), in the online world people were more likely to express "[t]he most unorthodox, imaginative, and subversive ideas" and doubt and ridicule all forms of authority (2). Such communicative messages and norms could be found in the proliferation of bulletin-board systems and blogs that emerged in the early years of the Internet. Finally, recent studies (for example, Chin 2011) indicate that some people in China prefer to interact and build relationships in online communities, as they find them freer and less constrained by social and cultural expectations.

The free flow of ideas on the Chinese Internet, however, did not go unchallenged by the Chinese government. Fearing that an uncontrolled Internet would lead to greater dissent and social uprising, China's government developed technology to monitor and censor the Internet. In 2003 the system of controls known in common parlance as the "Great Firewall of China" was launched (Canaves 2011). In subsequent years, as such Western-based social media technologies as YouTube, Twitter, and Facebook were developed, and used as a forum for expression and communicating messages of political dissent (for example, Iranian protests against the results of the Presidential election in 2009), China took the position that if a company would not allow the government to monitor and censor its content, it would be blocked. And during the lead-up to the twentieth anniversary of the Tiananmen Square massacre in June 2009 and the sixtieth anniversary of the establishment of the PRC in October of that same year, the government permanently blocked access to Facebook, Twitter, and YouTube.

Censorship, however, did not mean the end to China's active, creative, and authority-challenging online community. When Western-based media were shutdown, Chinese entrepreneurs and companies developed technologies that were not blocked, and were allowed to grow. For instance, the Chinese alternative to YouTube is YouKu; Sina Weibo has replaced Twitter (Talbot 2010). And since its launch in 2011, WeChat has quickly become China's version of Facebook.

The origins of WeChat can be traced to an earlier IM (Instant Messaging) technology, QQ, launched in 1999 by China's Tencent Company—the company that later developed WeChat. Like most IM applications of the time, QQ was a computer-based application for users connected to the Internet. Over time the technology was developed and then adapted for use on cell phones—as users could exchange messages using IM and/or email formats.

In 2011 Tencent Company launched WeChat. This technology was designed primarily for cell phones (Hou 2014). (WeChat is a free, downloadable app that generates most of its revenues from businesses that pay for a "public profile" used for commercial product promotion.) On a cell phone a user can select from a menu of four items: Chat, Contacts, Discover, or Me. (When a phone is set to a different language the layout is the same but the text changes; for example, in Chinese these four items appear as *Liaotian, Tong xun lu, Faxian*, or *Wo de sheding*.) When using Chat a user may send a text-based IM to other users on a contact list, much as they could do on QQ. On a smartphone, however, a WeChat user may "accent" a message by inserting one or more "emoji" icons. (Emoji is a term that refers to the ideograms used in text messages; it is Japanese in origin—similar to emoticon—and the term is used by speakers of a variety of languages across Asia, including English, Japanese, Cantonese, and Mandarin Chinese.) These range from such traditional images as smiley, winking, or sad faces, to more creative images that a user may download, and are often animated. (Most emoji are free; some can be downloaded for a small fee.) Examples of animated, downloadable emoji are shown in Figure 5.1: Garfield the Cat, Captain America, Molang the cute Bunny— from Korea, and Chinese characters such as 友—meaning "friend."

A second popular Chat function, and not available prior to WeChat, is the ability to send short (less than one minute) audio files to another

Figure 5.1. Emoji. Screen shot published with the knowledge and agreement of the WeChat users featured.

person, mimicking telephone calls. One often sees WeChat users speaking into the bottom of a cell phone where the microphone is located, and then moving it to the ear in order to hear an audio message sent in reply. Finally, users may create a user-defined "group chat" for sharing messages, pictures, and contact information. Some female users call this kind of WeChat group a *"guimi qun,"* 闺蜜群 or "girlfriends' group." (This term is gendered and identified with females; when a male joins such a group he is called a *"nan guimi,"* or "male girlfriend group member.") Thus, WeChat allows users to develop private networks among core friends (Chen 2013). Once you create a group it can be accessed repeatedly, so that multi-party interactions are more easily facilitated.

The "Me" menu items allow users to modify personal settings, such as inserting a picture or icon, changing a user name or password, or privacy settings. For instance, a user may choose to make a "Blocked list" or "hide" Moments postings (described below) from all but select contacts.

The "Contacts" item is similar to a contact list on a cell phone or social media application. Users may add or delete users from this list.

Perhaps the most interesting and novel menu item is the "Discover" tab. In addition to options for playing games or manually adding contacts—similar to other applications—users may discover other contacts in creative ways. One is to generate and exchange a QR code (Quick Response code) on a cell phone. When a QR code is scanned into another person's cell phone, contact information is exchanged. Or, one may "shake" a phone. When this option is selected, and the user shakes the phone, a message pops up showing the nearest person who is simultaneously shaking a phone and looking for contacts. It is then up to the users to decide if they want to make contact. (This function is more often used by males than females, and among those interested in a "hook-up" [Li 2014].)

The most popular Discover menu option is called "Moments." When selected, users see pictures and short messages posted by others on the contact list. Similar to Facebook, Moments is temporally ordered—the most recent items appear on top—and is a constantly changing screen of user generated pictures, comments, and responses. It is also a one-to-many forum for the exchange of personal experiences, news, items of interest, and commentary on everyday life. However, as will be discussed below, Moments postings do exhibit recognizable patterns.

In sum, while it may be argued that WeChat is popular across China because it is an unblocked alternative to Facebook, this explains its appeal only in part. The features described above show that it is a versatile social media platform for both one-to-one and one-to-many communication. In places such as Hong Kong or Macau, where Facebook is not blocked, many people have both WeChat and Facebook accounts; based on our observations and interviews we find WeChat is used more frequently as it offers more functions that connect users and/or facilitate social interaction. They may prefer to use WeChat as it has more functions than Facebook. And in a few short years WeChat has become one of the most popular social media platforms in China with more than 400 million monthly active users worldwide (Hong 2014).

METHODS

We now present data from a study that examines how and why university students use WeChat. Data were collected from 2013 to 2014 by ten graduate students at the University of Macau, located in Macao, a small Special Administrative Region (SAR) of China. Macao has a rich cultural tradition as it was administered by Portugal for four hundred years, and has served as an important meeting place between East (China) and West (Europe) (Hao 2011) (Guan and Sandel 2015). Since the handover of Macao from Portuguese colonial rule to China in 1999, and the opening of the casino concessions, Macao's population and economy have experienced rapid growth. Travel between the Chinese mainland and Macao has expanded with increases in the numbers of tourists, students, and workers. However, unlike mainland China where Facebook, Google, and other popular Internet sites are blocked, in Macao these sites are not blocked. Despite the availability of these sites, WeChat is used by many students, especially those who come from other parts of China. One possible reason for this prominent use may be that through WeChat they are able to keep in contact with friends and family members (Wu 2014). Macao, therefore, is a fruitful place for studying WeChat use.

For this study data were collected in the form of personal observations and screen shots of WeChat posts, and in-depth interviews with select users about their practices and the meanings they attach to these practices. The students and professor added each other to their WeChat contact list for the purpose of this research, and permission was granted to take screen shots. To protect participants' identities, students used screen names that differed from their real names. Finally, messages and images posted by others not in the research group (for example, the "teacher" whose messages are discussed below) were modified such that all identifying information was removed.

Our focus in this chapter is limited to the Chat and Moments functions of WeChat. These functions were used most frequently by participants in our research. Thus, we present exemplar screen shots and interpretive comments on these shots. We conducted in-depth interviews with a number of participants about how social media use in general impacts personal and social relationships, but did not include these data in this analysis. Finally, we did not study how people use WeChat's Audio file function, Shake, or the creation of WeChat Groups.

WECHAT AS CONTEXT

As claimed by Seargeant and Tagg (2014), communities that are linked through social media are not bound by time and space. The context for analyzing and understanding the code of WeChat is not a physical place, but the situations that arise when users communicate with others through and with WeChat. Therefore, while data were collected in one physically bounded place, Macao, and by a select group of people—students at the University of Macau—we collected data based on students' relational contacts maintained and facilitated through WeChat. Following Fitch's approach we looked specifically for situations when a WeChat message or posting prompted a reflection or comment on a rule of social interaction (Fitch 1999). We use WeChat screen shots to illustrate a social rule and then provide an analysis based on the interpretive comments gathered from participants. Based upon our own record of screen shots, and observations of how students use WeChat in daily life (for example, what they do when sitting in class, riding the bus), we found that the Chat and Moments functions were used most frequently. Thus, in the following we present data from these two types of communication.

ONE-TO-ONE COMMUNICATION

WeChat interactions vary based upon the number of interactants. Seargeant and Tagg (2014) suggest that "one-to-one" communication is

Figure 5.2. Chat 1. Screen shot published with the knowledge and agreement of the WeChat users featured.

a prominent way to use social media. By using the Chat function users can establish contacts with each other, and send individualized messages. This we see in the following:

The three messages presented in Figure 5.2 illustrate three interactional moves. As evidenced in the first message "I've accepted your friend request," one must first accept a request to initiate one-to-one interaction. In the second, posted eleven days later, a standard opening is offered, "Good afternoon Professor," followed by the question, "you are not in the office?" The third message offers the potential reason for the professor's absence, "Are you busy now?" and a potential solution, "Maybe I should come next day?" Notice how the text is interspersed with emoji—two smiley faces and one inquisitive face. These soften the force of the implied question: Did the professor forget the appointment? We also observe the use of indirect and politeness strategies in the language: "it seems that you are not in the office?" "Are you *busy* now?" Such language provides the tardy professor face saving reasons for being unavailable.

Figure 5.3. Chat 2. Screen shot published with the knowledge and agreement of the WeChat users featured.

In our study, we found WeChat users frequently included emoji embedded within their text messages. As noted above, most (for example, smiley face, frowning face) are provided by the service for free; users may also download animated emoji. Consider the following example:

This message was sent by a recently graduated student, Ban, who moved to a city in China where he began a job as a teacher. The second message on this screen shot begins with a surprise/happy face emoji (mouth covered by a hand) and the text, "ready to enjoy the feeling of being a teacher." The professor responds with an affirming message, "Glad you're working and a teacher." Ban responds with the explanation that he will do this job for now as he has no other alternatives. The next turn is a closing move, "Okay, keep in touch." The final turn by Ban is not a closing text, but an animated emoji: a figure with thumbs up. Mimicking the nonverbal expression of thumbs up, the emoji virtually extends a hand-and-thumbs-up gesture across the physical space that separates the two interactants. It shows how social media allows users to use resources at hand—text and emoji—to share not only information, but a nonverbal gesture and its associated emotional impact.

MOMENTS, ONE-TO-MANY: ACCEPTABLE POSTS

We now turn our attention to a discussion of how users employ the "Moments" function of WeChat for one-to-many communication. While we did not keep an exact accounting of how much time users spend on Chat versus Moments functions, we observed that users spend nearly equal time on both. For instance, when riding the bus a person may first carry on a Chat conversation with another person. Then, between turns, or after the Chat has ended, the person will select "Moments" and then scroll through the pictures and comments of that section. In the following we first discuss the most common types of postings made to Moments; this is followed by a discussion and interpretation of screen shots that demonstrate interactional dilemmas.

When communicating one-to-many, WeChat users post pictures and text through the "Moments" function. Food shots are common, as we see in the following:

The text reads (in Chinese—first author's translation): "[I] have discovered New Mainland [name of a restaurant]. Eat eat eat without stopping!!" Below the text is a composite picture of eight dishes and one shot of a drink.

Another common theme is to post travel pictures. This user, screen name "5-4," went on a trip to Taiwan. (Taiwan is a popular destination for

Figure 5.4.   **Moments One. Screen shot published with the knowledge and agreement of the WeChat users featured.**

mainland tourists.) The top picture is a "selfie" of the user's face covered by a scarf in order to protect herself from the wind, and accompanied by the text: "I am at Kenting with the weather . . . wind." The bottom is a photo of a temple procession, a common sight in Taiwan, but not in China. The focal figure is a character wearing the clothes of the Chinese deity, "Third prince," also known in Taiwan as "Tim-tau." 5-4 wrote: "I always wanted to see Tim-tau."

A third theme that emerges in many of the postings is to show participants doing interesting things jointly, featuring images of several people together.

Hazel posted pictures of herself and a friend who went to the city of Guangzhou (Canton). They are wearing colorful clothing, large sunglasses, and make-up. By their poses we see them physically close to each

五四 ✨
我在垦丁天气...疯 😜

30 days ago

五四 ✨
一直想看的阵头。

Figure 5.5. Moments Two. Screen shot published with the knowledge and agreement of the WeChat users featured.

other and smiling. We can infer a close friendship and happy mood. The text above these pictures reads:

> [We are] high [excited] after showing our colorful selves, and taking advantage of a moment when the driver did not pay attention, we were successful to get into a taxi. Because we are so colorful, when walking on the streets of Guangzhou, we are excited and pleased that people asked to have our picture taken with them.

Underneath the picture is a heart, indicating "likes" by Phoebe and Cherry. Cherry wrote the comment, "I am also in Guangzhou ! ! Feel that it is so much fun!" Hazel replied, "So much fun. Everyone is in Guangzhou (smiley face emoticon)."

The above Moments are examples of what users most often post on WeChat: Food, travel, and exciting and fun moments. All are considered post-worthy moments and preferred topics for sharing in one-to-many

Figure 5.6. Moments Three. Screen shot published with the knowledge and agreement of the WeChat users featured.

communication. They demonstrate a Chinese belief, captured by a saying in Taiwan: "If you find something good to eat, then report it widely to everyone" (Sandel 2015). It implies that if you find or experience something good, interesting, exciting, you then should share it with others. WeChat's one-to-many Moments function serves as a communicative tool for sharing what is considered good, interesting, and exciting news with others. Our examples illustrate that food and travel count as "news items" to share with others.

ONE-TO-MANY: PROBLEMATIC POSTS

In this next section we examine WeChat posts that became problematic. From previous work in ethnomethodology by Garfinkel, we know that violations help us see what rules interactants operate by in social situations (Garfinkel 1964). Such messages help us more clearly understand emergent rules for WeChat communication.

When talking with university students about what they found problematic with WeChat messages, one issue was how to respond to requests and/or messages from current and former teachers. Phoebe explained this in the following comment (this and following quotes were written in English and quoted verbatim):

> I really felt pressure to post "like" if my teachers post something on the Moments at the beginning when I use WeChat. Actually, I did feel pressure or nervous if a teacher try to follow me when I began to use WeChat. Because I used to think WeChat is a social tool for peers rather than elders. In Chinese students' mind, there is a clear bound[ary] which limits the relationships of students and teachers. Especially, when one of my high school teacher, who hadn't kept touch with me for several years before she followed my WeChat and asked me to post "like" in her posts since she participated in a commercial promotion. If she collected 30 "likes" she would have opportunity to win some prizes, such as a juicer extractor or a gold necklace or something like that. And she even sent me messages as: "Please help teacher to gather 'likes,' the teacher then can get the prize." I felt embarrassed and had to post "like."

This student's explanation demonstrates a number of interactional dilemmas. When Phoebe's former teacher initially requested her to "follow" on WeChat, and join her contact list, Phoebe felt conflicted: She considered WeChat to be "a social tool for peers rather than elders." WeChat is both designed as and considered a space, or community, for relational peers. The app does not create a hierarchically structured community, providing designations contrasting social roles such as teacher and student. Therefore, by accepting the teacher's request to "follow" her posts, Phoebe became aware that her posts not only were being sent to her social peers, but also to her teacher. Her reaction, feeling "nervous," may have been due to the change from a hierarchal to peer status since typically a person such as a teacher should be treated with respect, different from how one would treat a peer. This may be based upon a belief that there is a "bound[ary] which limits the relationships of students and teachers."

Another troublesome issue was the content and intent of the messages sent by her teacher. Users are encouraged to participate in this by posting commercial messages on their "Moments." If others then respond with a "like," the person who posts the message can receive a prize: "a juice extractor or a gold necklace or something like that." With the aim of "winning" a prize, this teacher sent messages to Phoebe asking her for "help." In response, Phoebe felt "embarrassed" and replied with a "like." It is unclear if the feeling of embarrassment was for the teacher or Phoebe

Figure 5.7. Moments Four. Screen shot published with the knowledge and agreement of the WeChat users featured.

herself. Most likely it is a combination of the two; it may also be due to the realization that her responses are visible to other classmates and WeChat "friends" of the teacher, meaning that she was aware of the one-to-many nature of Moments.

Interactional dilemmas may also occur in the context of business relationships. Ms. Lee worked as a trainee for an international company in China. Lee added both her director and department manager to her contacts list, thus providing access to her Moment's messages (and in this case access to their Moment's postings). Consider the following example:

The picture was taken by the director from inside her car when leaving the company parking lot at the beginning of her journey home. The accompanying text reads: "At this time when getting off work it is still light outside. Summer is coming (smiley face, sun). [I] hope that there is not a traffic jam, and can quickly see [my] treasure [child] (heart)." Lee, whose icon is the bear, quickly responded with a "like" indicated by a heart and her icon. Others replied with brief messages about the journey and the director's expected reunion with her child, accompanied by appropriate emoticons—two smiling faces and one sun. When later reflecting on this post, Lee commented:

> As for the pressure to press "like" when it comes to my boss or teachers, I do have the same feeling as your students. As the attached WeChat screen shots show, when I was a trainee in [company], [Name] and [Name] are the director and department manager, I would always press "like" when they posted pictures, to show my respect or care or something else.

We see that the hierarchical relationship that Lee had with her superiors in the offline, business world impacted how she behaved in the online, WeChat social world. Furthermore, this was felt as a form of "pressure" in that Lee was obligated to respond with a "like" in order to communicate her "respect or care."

The above sentiments, however, were not shared by all students. One student commented:

> As for the problems of hierarchy in WeChat use, I don't feel any pressure to click "like" (that means if I press like just when I like). The same opinion is held by my 2 roommates. But most of us feel a bit embarrassed to expose our personal life or feeling on MOMENTS if our boss or teachers are watching it.

This student claimed that she does not feel pressure to respond by pressing "like" when the poster is a superior, such as a boss or teacher. However, she and others are aware that when posting to Moments the communication is one-to-many, and that if a boss or teacher sees it, messages which are too personal may cause a feeling of embarrassment.

Another student similarly said that he does not feel he has to press "like" to Moments postings made by his professors. Yet he is aware that WeChat posts can be problematic and has developed his own rules for avoiding them:

> I think sometimes I am afraid some strangers or acquaintance want me to add them as my friends in WeChat. For one reason, I am not feel free to share all kinds of information in Friend Circle, although there's a function make these people in a group and offer them a limitation on specific information. For another reason, you may ignore that each people could copy your portrait image and your name which they could pretend to be you and cheat your friends' money.

One rule is to not add strangers or people who are not relationally close, or categorized as an "acquaintance" as friends in WeChat. Another is to be careful what information he shares with people—even those who are in his "Friend Circle." Finally, he may be aware that some people may use WeChat to defraud others by using your portrait and name, pretend to have your identity, and then try to cheat friends from their money.

CHANGING AND ADAPTING

In this last section we discuss changes in the ways people use the Moments function of WeChat. Some of the above comments allude to this. For instance, there is a growing awareness that WeChat is not always safe, and that you need to be careful when adding friends and/or replying to requests. Yet this is not the only change.

Phoebe, who above said that she initially felt troubled when her teacher asked her to post "like" to a WeChat promotion, explained that she now sees her teacher's posts differently:

> But as time goes by, I realized that it is good and usual to contact your teacher with new media. And I get used to looking at my teachers' posts, and treating them normally. If the content is meaningful and interesting, I feel like to post a "like" in it. . . . Not only me, but some of my friends do that.

Now that she is more accustomed to interacting with her teacher through WeChat, she sees her teacher not in the role of hierarchically superior, but as "normal," more as she would treat a peer. If she finds her teacher's post to be interesting, she will respond with a "like," but does not feel compelled to do so. This is something that she observes her friends also doing, and demonstrated this with screen shots posted by her teacher that several students responded to with "like."

Changes to students' responses to teachers' postings, however, have continued. Phoebe then commented: "But one interesting situation is that, most of my classmates are tend to ignore teachers' posts now, since the times of contacting is less than before and we are all accustomed to teachers' posts." She illustrated this with two screen shots made by the same teacher. One received just one "like" from a student. The other, which Phoebe claimed best illustrated her point, was one that neither Phoebe nor her WeChat friends responded to. (As indicated by the teacher's comment, at least one other person—not on Phoebe's contact list and hidden from her—posted a response.). Following this unknown comment, the teacher then responded, as we see in the following:

The teacher wrote at the top: "[My] daughter spent the whole day at the swimming pool and when she came back, her prize was to go to Tangren Street and eat sweet snacks. Go!" Below this are pictures of her daughter and the snacks that she ate. However, neither Phoebe nor the classmates on her contact list responded to this with a "like" or made a comment.

Figure 5.8. Moments Five. Screen shot published with the knowledge and agreement of the WeChat users featured.

A day later, the teacher responded to another person's posting with the comment: "So dark. [Name] little sister now is truly a dark girl. But you should go to school and learn how to swim. Do not waste this opportunity." This last comment is interesting on a number of levels. For one, she is commenting on her daughter's skin tan that has turned her "dark," and—by Chinese standards of beauty—tanned skin is unattractive and undesirable. Thus, her daughter "sacrificed" some of her beauty when learning how to swim. But in the next sentence the teacher takes on the voice of the teacher and speaks directly to her students. This is evident in the use of the indexical pronoun "you" that is pointing to the students— her presumed audience: "But *you* should go to school and learn how to swim." Her students should not let the fear of skin turning dark stop them from learning how to swim, and using the swimming pool that their school has provided.

It appears that the teacher is using WeChat to communicate hierarchically to her students, just as she would in face-to-face interaction. Perhaps the "absence" of likes is interpreted by the teacher as an expression of their displeasure, or some kind of dislike about the post. That is, the lack of likes also sends a message, and the teacher has learned to respond (and perhaps use the app as well) by re-asserting her voice as teacher, and someone who is due proper respect. Nevertheless, as Phoebe explained, many students chose to ignore this "teacherly" voice, did not reply, and this conversation came to a close.

CONCLUSION

This chapter began with the observation that cell phone use in China has grown rapidly, and this device impacts behavior and daily interactions. By focusing on how university students use one mobile app called WeChat, we have found that users negotiate their preferences to engage in one-to-one or one-to-many communication. Significantly two different tools enable the movement from each of these types of communication. WeChats' "Chats" function allows users to send instant text or audio file messages to one other person. The "Moments" function facilitates one-to-many communication, as a picture and accompanying text can be sent out, viewed, and responded to by many users.

What is of greater interest in this chapter, however, is to describe and interpret the rules of use or communicative codes that participants implicitly and explicitly follow when communicating with others via WeChat. These data demonstrate that when engaging in Chat, or one-to-one communication, users enhance the dramatic and emotive content of their messages with emoji. This resonates with Walther's earlier studies of computer-mediated-communication (CMC), that claim CMC is not inferior to face-to-face, as participants use resources that are at hand (for example, emoji and/or emoticons) to convey affect and emotion (Walther 1992, 2012).

The Moments function of WeChat, which enables one-to-many communication, demonstrates other rules of interaction. One emerges when observing the content of most posts: users share pictures and texts of "fun" activities, such as food, travel photos, or happy moments. This resonates with the Chinese folk concept of wanting to "share good news with others" (Sandel 2015). The second rule is based on the understanding that WeChat as a social environment is not always safe. It is understood that when an acquaintance or stranger makes a request to be added as a contact, this person is a potential danger, and may try to engage in fraud or other deceptive activities. Thus, users may be more cautious when considering who to add as a contact.

A third rule guides how students interact with hierarchical superiors, such as teachers or bosses. The Chinese social world is hierarchically structured (Chang 2010) (Gao and Ting-Toomey 1998) and when subordinates talk to superiors (e.g., child to parent), subordinates should show proper deference (Sandel 2004). Our data indicate that students feel a higher degree of "nervousness" when responding to messages and/or requests from teachers or bosses. Students may feel pressured to "like" a post made by a teacher, not because they truly like the post, but because they want to show the proper level of respect. Yet what is most interesting is to see that this interactional rule appears to be changing. With the passage of time, and as students become more accustomed to seeing a teacher's post, they begin to treat the message as a "normal" post by a classmate or friend. Thus, they may ignore the post, or feel inclined to press "like" because of the content of the message, and not the position of the sender. In this way we are seeing that—despite the efforts of some (e.g., the teacher) to assert an authoritative voice—WeChat is impacting the perception of a social hierarchy, and leading to a social world that is more egalitarian and less hierarchical.

In sum, we see that WeChat use not only reflects communicative codes from the offline world (e.g., hierarchical deference), but may also be changing it. WeChat may be a communicative platform that "flattens" the social structure. It thus exposes users to a world that is more open to dangers and cheats, but also brings superiors' everyday activities and banal moments into subordinates' worlds. We see in these posts and interpretive comments the struggle to define this new community of practice, much as we saw in Fitch's earlier observations of a Listserv (Fitch 1999). This also hearkens back to Hymes's work among the Wishram Chinook who had special rules for defining and interpreting what counts as language—infants' vocalizations. WeChat users in greater China are developing new rules to decide if a teacher's posting is worthy of a "like," or is a posting from a "normal" (social peer) and can be ignored if the content is not interesting and worthy of comment.

NOTES

1. Publisher's Note: The screen shots used as supplemental research in this text are included with the participants' knowledge and agreement that these screen shots would be used in a later publication.

2. Acknowledgments: We would like to give special thanks to the following students at the University of Macau for help with the conceptualization, data collection, and interpretation of this study: Mengjie Chen, Agnes Choi, Shengmei Li, Peiyuan Lin, Jianxia Wan, Si Wu, Li Jiamin, and Li Zhong. We also thank Richard Fitzgerald, also of the University of Macau, for his helpful advice and comments.

REFERENCES

Baron, Naomi S. "See You Online: Gender Issues in College Student Use of Instant Messaging." *Journal of Language and Social Psychology* 23, no. 4 (2004): 397–423.

Canaves, Sky. "China's Social Networking Problem." *Spectrum, IEEE* 48, no. 6 (June 2011): 74–77.

Chang, Hui-Ching. *Clever, Creative, Modest: The Chinese Language Practice.* Shanghai, China: Shanghai Foreign Language Education Press, 2010.

Chen, Yongdong. "Wechat and Weibo: Complement Not Substitutes [In Chinese]." *News and Writing* 4 (2013): 31–33.

Chin, Yann-Ling. "'Platonic Relationships' in China's Online Social Milieu: A Lubricant for Banal Everyday Life." *Chinese Journal of Communication* 4, no. 4 (2011): 400–416.

Fitch, Kristine L. "Pillow Talk?" *Research on Language and Social Interaction* 32, no. 1–2 (1999): 41–50.

Gao, Ge, and Stella Ting-Toomey. *Communicating Effectively with the Chinese.* Thousand Oaks, CA: Sage, 1998.

Garfinkel, Harold. "Studies of the Routine Grounds of Everyday Activities." *Social Problems* 11, no. 3 (1964): 225–250.

Guan, Xin, and Todd L. Sandel. "The Acculturation and Identity of New Immigrant Youth in Macao." *China Media Research* 11, no. 1 (2015): 112–124.

Hao, Zhidong. *Macau: History and society.* Hong Kong: Hong Kong University Press, 2011.

Homsey, Dini, and Todd L. Sandel. "The Code of Food and Tradition: Exploring a Lebanese (American) Speech Code in Practice in Flatland." *Journal of Intercultural Communication Research* 41, no. 1 (2012): 59–80.

Hong, Kaylene. "WeChat Climbs to 438 Million Monthly Active Users." *The Next Web.* August 13, 2014. Accessed January 14, 2015. http://thenextweb.com/apps/2014/08/13/wechat-climbs-to-438-million-monthly-active-users-closing-in-on-whatsapps-500-million/.

Hou, E. "Government Wechat, Expired 'Old Ticket' [In Chinese]." *China Media Report Overseas* 10, no. 1 (2014): 1–7.

Hymes, Dell. *Foundations in Sociolinguistics: An Ethnographic Approach.* Philadelphia, PA: University of Pennsylvania Press, 1974.

Lee, Hung-Tien. "Songshan xian tong che, ditou zu ye kandedao!: Songjiang Nanjing zhan she da xing yin dao di tie [Songshan line, even the head lowered tribe can see!: Songjiang Nanjing station has set up a very large passenger direction sign." *NOWnews*, 11 (2014): 11. Accessed November 12, 2014. http://www.nownews.com/n/2014/11/11/1500924.

Li, Zhuo. *Weixin dui qingnian qunti ren ji guanxi yingxiang de yanjiu [The study of WeChat's influence on young people's relationships].* Masters Thesis. Inner Mongolia University, 2014.

Milburn, Trudy. "Speech Community: Reflections upon Communication." In *Communication Yearbook*, edited by Pamela J. Kalbfleisch, 28, 411–441. Lawrence Erlbaum/ICA, 2004.

Philipsen, Gerry. *Speaking Culturally: Explorations in Social Communication.* Albany, NY: State University of New York Press, 1992.

Philipsen, Gerry, Lisa M. Coutu, and Patricia Covarrubias. "Speech Codes Theory: Restatement, Revisions, and Response to Criticisms." In *Theorizing About Intercultural Communication*, edited by William Gudykunst, 55–68. Thousand Oaks, CA: Sage, 2005.

Sandel, Todd L. "Narrated Relationships: Mothers-in-Law and Daughters-in-Law Justifying Conflicts in Taiwan's Chhan-chng." *Research on Language and Social Interaction* 37, no. 3 (2004): 365–398.

Sandel, Todd L. *Brides on Sale: Taiwanese Cross Border Marriages in a Globalizing Asia*. New York: Peter Lang, 2015.

Seargeant, Philip, and Caroline Tagg. "Introduction: The Language of Social Media." In *The Language of Social Media: Identity and Community on the Internet*, edited by Philip Seargeant and Caroline Tagg, 1–20. New York: Palgrave MacMillan, 2014.

Seargeant, Philip, Caroline Tagg, and Wipapan Ngampramuan. "Language Choice and Addressivity Strategies in Thai-English Social Network Interactions." *Journal of Sociolinguistics* 16, no. 4 (2012): 510–531.

Talbot, David. "China's Internet Paradox." *MIT Technology Review*. (April 14, 2010). Accessed February 6, 2015 and January 9, 2015. http://www.technologyreview.com/featuredstory/418448/chinas-internet-paradox/.

Walther, Joseph B. "Interpersonal Effects in Computer-Mediated Interaction: A Relational Perspective." *Communication Research* 19, no. 1 (1992): 52–90.

Walther, Joseph B. "Interaction through Technological Lenses: Computer-Mediated Communication and Language." *Journal of Language and Social Psychology* 31, no. 4 (2012): 397–414.

Wu, Si. *I am Not Fighting Alone: The Impact of Social Media on the Cross-Border Study Experience of Mainland Students in Macau*. Macao, SAR: Master's thesis, University of Macau, 2014.

Yang, Guobin. *The Power of the Internet in China*. New York: Columbia University Press, 2009.

III

INTERCULTURAL INTEGRATION
Trudy Milburn

In order to help developers and designers learn more about the people using their new digital devices, researchers in this section address the question, "which cultural premises are being evoked when people interact with a specific digital medium in a particular situation?" This aligns with Murray's (2012) comment about innovative design, where she suggests that "we look beyond received opinion and familiar solutions to identify the deeper, cultural connections" (40).

Research subjects in the previous two sections have been referred to as users, actors, participants, speakers, and members, typically based on a methodological preference. In addition, each chapter includes reports about interaction based on participants' own role-specific labels, such as board member, student, trainer, driver, friend, etc. One of the ways LSR scholars attend to cultural identity is to listen for the terms used to describe people. As some authors have mentioned, each participant designation implicates a person into role obligations. These obligations include the ways one can and should relate to other participants as well as how one can and should use digital media when interacting. For instance, as the introduction pointed out, even when employing the term "users," we understand the rights, duties, and obligations of a person acting with a digital medium in a particular way.

We may ask of terms that describe digital media users: Is the "user" like the "culturally preferred model person, the individual" (Scollo 2011, 13)? To tease this apart, researchers should look closely at the way digital media are designed. Is the user assumed to be an individual or a group of people (as Peters questions in chapter 3)? If so, what impact do these

assumptions have for designers? Additionally, do specific use-cases reinforce assumed or preferred models of identity?

By learning more about the cultural premises of those utilizing digital
media to interact, LSR seeks to learn more about how users present themselves as specific kinds of people, within particular situations.

THEORETICAL PREMISES

Murray (2012) describes the importance of creating systems that feature
persons as agents (or with agency). Personhood then, is comprised of
both the person with an identity (or multiple identities) as well as the
role positions that person has with other persons (i.e. teacher—student,
etc.). These premises suggest a move from individual actions and interpersonal relationships to infer a type of user personhood or persona. This
section draws the full implications from a single person with agency who
"mak[es] something happen in a dynamically responsive world" (Murray
2012, 100) to a person embedded in a social world, with all the webs of
connections that implies.

Thus far, the chapters have explored specific actions and interactions
that take place and have suggested some cultural implications of those
actions. The chapters in this next section specifically address the larger,
symbolic boundaries of culture and its importance for the design and
development of digital media by following Carbaugh's (1994) description
of persons as "agents-in-society" (163). In order to understand users, we
adhere to Carbaugh's methodological suggestion to attend "carefully to
the cultural structuring of personhood in interactional processes" (162).
While agents-in-society can be considered "users" who purposefully act
and form relationships through interaction, Carbaugh's (1994) research
reveals some particular American assumptions about personhood, such
as the way individual "selves" are positioned relative to others individual
"selves."

Both chapters in this section presume that persons are constituted
through their "discursive activity" (or their "expressive practices that
make available particular positions for persons to take up and address
and with which to hear others taking up and addressing" [Carbaugh
1994, 164]). Furthermore, they explicitly take up the question that has
been implicit within the previous two sections: As users engage or act in
ways identified above, what identities are they enacting?

Scollo describes the importance of discovering the "connection of
communication to models of personhood, or who and what the model
person is conceived to be in different cultures" (2011, 7). In her review

of cultural discourse theory (CDT), Scollo describes the ongoing investigations into discourses of personhood and relationships within different countries and cultures. Scollo admits that different cultures may emphasize one way of communicating over others (perhaps focusing on emotions over actions, for example—and in fact, we see some of this in chapter 6).

While the creation of personas has long been part of typical design process, it has generally been done to conceptualize prototypical users. Such processes limit potential users to somewhat static categories of people who will use the product. What is illustrated in the following chapters is a means to strengthen the use of persona building in design. In concert with typical design practice, the significance of this way of working is to create a persona in advance of new design, to target a particular audience of users (much the way market analysis has worked), but to do so by envisioning a particular prototypical user in terms of communication and identity. The features of culture that help to construct these prototypical users are extremely important to consider if our products and practices are used globally.

In the two chapters in this section, Poutiainen and Mackenzie and Wallace describe research that helps us move analytically from an investigation of users interacting directly with digital media to interpretations about cultural premises. These chapters help to illustrate the way research can be used to understand cultural differences that may be important for the design, development, and use of digital media.

Poutiainen illustrates the way Finnish cultural identity has been both reinforced as well as modified by their mobile phone use. This chapter provides evidence for ways that historic premises about culture are prominently referred to and reacted against when considering communication through digital media. Through news articles and interviews on mobile phone usage over a ten-year period, Poutiainen elaborates our understanding of the ties between Finnish cultural identity and technology, and the ways these may be shifting.

Mackenzie and Wallace describe the way military students' cultural experiences are employed as pedagogical resources for their design and continued refinement of an online course. Their chapter describes their process for incorporating students' contributions to a course-based wiki into the design of the next iteration of their course. Mackenzie and Wallace draw upon students' diverse reference to their experiences within culture-specific scenario within countries ranging from Singapore, South Korea, Italy, the Bahamas, and Canada to provide the most specific interactional examples they can for the military personnel who may find themselves in a similar situation.

WHAT ARE THE IMPLICATIONS FOR DESIGNERS?

This section provides strong evidence of the importance of attending to cultural premises when designing digital media. While Murray (2012) has asserted that "the designer should think about the core human needs served by the new artifact, survey the way those needs have been met across multiple platforms and then attempt to reimagine them as they might be served by the affordances of the emerging digital medium" (23), this section suggests that we move beyond the core human needs to more specific cultural premises for acting and relating.

In the two chapters in this section, the authors present cultural premises within on various situations. Here threads from the previous sections, such as communicative activities including teaching and learning, are combined with the creation and maintenance of social relationships. Because culture is enacted in different ways, the two chapters illustrate very different methods for understanding culture: one through direct reflection about experiences that were based in unfamiliar countries and one through in-country reflections about citizens' own practices as compared with those outside of their country.

Special forums devoted to the exploration of intercultural communication practices and new media illustrate growing interest in this area. For instance, in 2011, Shuter found limited published research about new media and intercultural communication research and subsequently recounts that his call for the special forum produced sixty submissions. The generation of scholarship in this area dovetails with the local strategies research based on investigations of cultural practices. By attending carefully to the specific ways that people interact, we can more clearly understand who they are as individual communicators, who they take their interlocutors to be, as well as underlying beliefs and values that can be said to be the basis of their cultural identities.

The chapters in this section help us to appreciate the ways that humans display, present, and to some extent, evaluate the quality of their interaction with others based, in part, in the displays of cultural identities in particular social scenes. Let us turn to these examples now.

REFERENCES

Carbaugh, Donal. "Personhood, Positioning and Cultural Pragmatics: American Dignity in Cross-Cultural Perspective." In *Communication Yearbook*, volume 17, edited by Stanley A. Deetz, 159–186. New York: Routledge, 1994.

Murray, Janet H. *Inventing the Medium: Principles of Interaction Design as a Cultural Practice*. Cambridge, MA: The MIT Press, 2012.

Scollo, Michelle. "Cultural Approaches to Discourse Analysis: A Theoretical and Methodological Conversation with Special Focus on Donal Carbaugh's Cultural Discourse Theory." *Journal of Multicultural Discourses* 6, no. 1 (2011): 1–32.

Shuter, Robert. "Introduction: New Media Across Cultures—Prospect and Promise." *Journal of International and Intercultural Communication* 4, no. 4 (2011): 241–245.

SIX

Myths about Finnishness

On Cultural Mobile Phone Discourses
Saila Poutiainen

Finnishness and technology are claimed to have a connection—technological developments are a part of the sense of Finnishness. This article
examines the ways in which Finnishness and mobile phone communication are united in talk and writing about mobile phone communication. This claim is supported by at least two arguments. Firstly, it has
been earlier acknowledged both in non-academic writings (for example,
Häikiö 1998, Mäenpää 2004, Wiio 1999) as well as in empirical research
(Vehviläinen 2002) that Finnishness and technology are somehow joint
projects for Finns. Secondly, due to Nokia's (past) reputation as a mobile
phone manufacturer, Finland is known world-wide as a country that
knows how to produce technology. Mobile phones, and the business
around them, have played a major role both in the Finnish economy as
well as in its gaining a reputation of being a technology-savvy country.
The fast growing information technology industry of the early 1990s was
described, for example, as "bringing Finland up from the forest, past the
European civilization, to something new and better" (Mäenpää 2004, 518).
Exploring Finnish cultural discourses about mobile phone communication, the question asked in this article is, how exactly do Finns talk about
the ways in which Finnishness and mobile phones are connected? I will
first sketch the theoretical framework of ethnography of communication,
and its stand on cultural and social identity and myths. As a part of that, I
will introduce *suomalaisuuspuhe* (Finnishness talk) as it is presented in the
existing academic literature. I will then move on to present data collected
in Finland, and the ways in which, in the data, Finnishness, *suomalaisu-*
uspuhe in particular, and mobile phone communication are united. I will
finish with discussing the cultural beliefs and values about being, social

relationships, or communication that are active in the talk about Finnishness and mobile phone communication.

IDENTITY, MYTHS, AND FINNISHNESS TALK

To study cultural, social, or national identities from the ethnography of communication perspective brings forward three working assumptions from which to look at identities as communication (Carbaugh 1996, 24). First of all, identities can be looked at as dimensions and outcomes of communication practices. By studying communication actions carefully, we can see who are the ones engaged in the action. It is in the social scenes in which we participate where our identity (or multiple identities) comes to the fore. For that to happen, an individual needs to know how to express that particular identity effectively and how to have it validated.

The second working assumption is that, just like communication, when identity is seen as created in interaction, each identity is seen as a set of communicative practices that are salient in some but not all social scenes. In other words, some social scenes are set for some identities, and not for others. When identity is seen as communication, then your own and others' sense of who you are depends on the actual conceptions as well as the conduct (or "doing") of identity in real social scenes (Carbaugh 1996, 25–26).

The third working assumption is that, just like communication, the everyday practices of social identities are also cultural accomplishments. Earlier ethnographic studies of communication describe, for example, gendered identity of a man and the ways in which to speak like one (Philipsen 1975), or "real Indianess" among Native Americans (Wieder and Pratt 1990).

Carbaugh further states that to locate the social and cultural identities in communication in particular social scenes, we should look at the symbols used and the meanings, intentional or not, that are salient when hearing or reading a symbol used (Carbaugh 1996, 83). To understand the dimensions and qualities of cultural identity, one should look at terms for identities as well as investigate their usage in various contexts, the styles and tones used, and the shared readings of any particular use. A way to construct, share, affirm, and negotiate an identity is to present it in a myth (Philipsen 1987, 250).

Myths, along with rituals and social dramas, are considered in ethnographic studies of communication as processual enactments in and for the community. Myth, according to Philipsen (1987, 251), is "a great symbolic narrative which holds together the imagination of a people and provides bases of harmonious thought and action." Public myths are resources or places to negotiate and discover the past and the present, the tension or the fit between community and individual, and to deal with desires and fears. Public myths are thus interdependent. Further, myths can be looked at as a displaying "culture's interpretive and rhetorical resources," revealing rich points to examine cultural symbols and cultural codes (Berry 1995, 46).

Apo (1996) suggests that particular ways of talking about Finnishness are also a myth. The myth of Finnishness, as described in detail in the following, was created in nineteenth-century Finland, during the nationalistic era. It tells a story for Finns about what it has been and is to be Finnish in Finland, what the Finns are by nature, what kind of a relationship Finns have with the environment, and what are the values and beliefs that have pulled Finns from the woods into civilization.

FINNISHNESS TALK

The meanings, functions, and versatility of Finnishness talk (literally translated from *suomalaisuuspuhe*) have been described by several scholars in different fields of academia (see, for example, Alasuutari 1998; Anttila 1993; Apo 1996 and 1998; Gordon, Komulainen, and Lempiäinen 2002; Lehtonen, Löytty, and Ruuska 2004; Leskinen 2005; Nieminen 2000; Peltonen 1998; Ruuska 1999 and 2002; Varpio 1999). Finnishness talk can be described as an example of discursive construction of national identity (for example, Wodak, de Cillia, Reisigl, and Liebhort 2009). Finnishness talk, in short, is the repeated and patterned way of talking about Finnishness among Finns themselves. Alasuutari (1998, 164, 169) claims that Finnishness talk is so familiar that every Finn recognizes it and is able to produce the talk. As an ethnographer of communication I have recognized the same as Ruuska (1999, 294): when asking Finns about Finnishness, typical Finnishness talk emerges rather quickly and strongly.

Finnishness talk is both negative and positive (Apo 1998). The negative Finnishness talk, which is much more dominant, is described by Apo (1996) as self-racism—a rather strong term, but deliberately so—to describe the hyper-critical attitude explicit in the talk. Some of the typical notions considered negative are that Finns emerged from the woods, they have low self-esteem, they are uncivilized or ignorant, non-Europeans, unable to interact or talk, slow, shy, and socially clumsy. The more positive Finnishness talk emphasizes the mythic and heroic Finn or the noble, god-loving, hard-working (*uuttera*), and law-abiding farmer.

The roots of Finnishness talk are found in nineteenth-century nationalism, in the years before Finland gained independence from Russia in 1917. Finland was under Russian power for about one hundred years, and before that, Finland was part of the Swedish monarchy for about five hundred years. During the era of nationalism, initiation and development of nationalistic thoughts was first something of an educated hobby for Swedish speaking *sivistyneistö* (intelligentsia). Later, Finnish nationalism developed into a movement led by middle-class academics, wealthy farmers, and rural clergy (Leskinen 2005), and the grassroots element of the nation became aware of the nationalist movement only in the late nineteenth century and early twentieth century (Alapuro 1994, 83).

Leskinen (2005) and Apo (1996) suggest that Finnishness talk is connected to *snellmannilainen ajattelu* (Snellmanian thinking). J. V. Snellman was a Finnish academic, writer, statesman, and important actor in the nationalist movement. According to Snellman's thinking, the nation needs to be aware of its national spirit (*kansallishenki*), create a strong sense of nationalism (*kansallistunto*), and unique national education (*sivitys*) (see, for example, Leskinen 2005). In Snellmanian thinking, continuous education or further attempts at civilization are preconditions for national spirit, and Snellman himself stated that *"pienen kansan turva on sivistyksessä"* (the safeguard of the small nation is in civilization). The main project in building the independent Finnish nation was to develop a nation that would feel united.

Apo (1996) has noted that in the definitions of Finnishness, both the "high-culture" of the European nations as well as the culture of the original Finnish folk are cherished. The nationalist project in the nineteenth century involved these two "cultures" and their union (Apo 1996, 15). On the one hand Finnish *sivistyneistö* had to rely on the Finnish speaking agrarian people as the resource and home for Finnishness, and on the other, the original rural Finns being committed to the goal stated by *sivistyneistö*. The shared goal by *sivistyneistö* and *kansa* was *isänmaan asia* (the matter of the fatherland). To achieve the goals of Finnish independence and a strong sense of Finnishness, both the Finnish speaking agrarian people and *sivistyneistö* had to pay a high price: A large part of *sivistyneistö*, for example, changed its first language from Swedish to Finnish even to the point where Swedish speaking parents changed the language they spoke to their children. The Finnish agrarian people committed to continuous and rapid changes, the main aim being to develop themselves into proper citizens. The negative *suomalaisuuspuhe* emerges from these changes: Finns were described as needing, for example, education, civilization, proper manners, and better hygiene. According to Apo (1996), the ways Finns, *kansa*, were described and talked about less than two hundred years ago are still active and observable in the current discourses about Finnishness.

Based on empirical research, Apo (1996, 18–22) provides three main qualities of Finnishness talk: totalizing, metonymy, and cultural evolutionism. By totalizing, the speaker uses expressions in which she or he makes poorly argued generalized claims about the whole Finnish nation and culture. Metonymy refers to expressions in which some small part of the country or the nation is believed to represent the whole country. The third quality, cultural evolutionism, points to an old scientific belief system from the nineteenth century, and it placed Finns below the so called "old culture" nations. Expressions in which Finns are described as underdeveloped, primitive, or archaic, are examples of an implicit belief in cultural evolutionism. Upon presenting these qualities, Apo nullifies the intellectual bases for them, and shows that the ways to talk about Finnishness or Finns are inconsistent with the present and historical reality.

Finnishness talk has its roots in the lived Finnish nationalistic history. The mythic form cultivates a sense of Finns as, for example, shy, hardworking, honest, modest, uneducated, and reluctant to communicate. These qualities provide bases for the cultural identity of the Finn, and thus position it as such. This myth is retold in research interviews and in media texts that formed the basis of this study.[1]

OBSERVATIONS ON FINNISHNESS TALK IN TALK ABOUT MOBILE PHONE COMMUNICATION

A Need for Education

Let me start with an example from the data. In a newspaper column published in 2005 and titled *Tolkutonta kännykkäkäyttäytymistä* (Insane mobile phone behavior), the writer stated that the mobile phone takes too much of the social space and attention, and that those using mobile phones are unaware of the proper social etiquette. Example 1:

> *Kännykästä on tullut puolijumala. Se on jatkuvasti käden ulottuvilla ja kuiskuttelemassa korvaan. Sen kutsuun vastataan auliisti–ihan älyttömissäkin paikoissa–välittämättä omasta seuralaisesta tai muista ihmisistä ympärillä. Nykyihminen näyttää olevan sellainen, että oma järki ei sano yhtään mitään käyttäytymis- ja kohteliaisuussäännöistä, kun kännykän kutsuun on vastattava. Kännykkäkansa pitäisi laittaa kännykkäkouluun oppimaan soveliaat kapulankäytön alkeet.* (By Anne Laurila in Länsi-Suomi, 260105[2])

> The mobile has become a semi-god. It is constantly at the reach of the hand and whispering to your ear. Its call is answered eagerly – even in insane locations – without caring [about] one's own companion or other people around. The modern human seems to be such that one's own sense doesn't say anything about rules for behavior and politeness, when one needs to answer the summons of the mobile phone. The mobile phone nation should be put into mobile phone school to learn suitable basics in the mobile use.

The writer refers to the mobile phone users as a mobile phone nation (*kännykkäkansa*). This nation needs further education to learn to behave suitably. The text introduces a repeated and widely recognized way of writing about Finnishness: Finns need to learn or need to be taught to communicate. The idea of learning and schooling is active, for example, in describing older people, who have been the last to adopt mobile phones. They have been offered opportunities to take part in *kännykkäkoulu* (mobile phone schools), which are introductory courses to mobile phoning arranged by, for example, group homes for the elderly or municipally-owned adult education centers. These days the mobile phone schools focus on learning the use of smartphones. Occasionally attendance of *kännykkäkoulu*, as in the example above, is recommended for those who do not know how to communicate or behave properly while using the mobile phone.

Another writer, in her letter to the editor, expressed misgivings about the increase in prices for landline phone calls. Higher prices in landline telephone use were expected since the number of landline subscriptions was decreasing, and the number of households having mobile phones as their only telephone device was increasing. The writer points out that some social groups such as students or people living alone cannot afford the higher prices. The writer is convinced that Helsinki Telephone Corporation (HPY) is pushing Finns to become mobile phone users. Example 2:

> *Onhan toki kännykät. Luulin pitkään, että HPY pyrkisi edes jotenkin kilpailemaan niiden kanssa ja yrittämään asemansa säilyttämistä juuri rauhallisten kotipuheluiden tuottajana. Turhaa luuloa, jo perusmaksun korotus kertoi selvin sanoin, että meistä tehdään väkisin kansaa, joka osaa vastata puhelimeen ainoastaan moi ja jatkaa puhelua parilla sanalla, okei, joo mä soitan taas, moi. Ja joku vielä väitti, että kännykät opettivat suomalaiset puhumaan.* (By Elisabet Aho, kirjailija, Vantaa in Helsingin Sanomat, 110499)

> Surely there are the mobile phones. I thought for a long time that HPY would even somehow compete with them and try to maintain its position as a producer of peaceful home-telephone calls. Quite mistaken, already the increase in the basic fee tells clear and loud that we are made by force a nation which can answer the phone only with *moi* (hi) and continue the phone call with few words, *okei, joo mä soitan taas, moi* (ok, yes, I will call later, bye). And somebody made the claim that the mobile phones taught Finns to talk.

The writer implies that mobile phone communication is short and about unimportant matters. This kind of communication is not difficult, meaningful, or skillful. The writer is worried about the Finnish nation, the ways in which the nation is treated by the telephone company, and the state of education of this nation. The last sentence in the excerpt has a tone of frustration and deep irony: The mobile phones were supposed to be good for Finnish communication, they were supposed to teach Finns to talk, but obviously, according to the writer, this has not happened. That is why the writer argues that mobile phones should not be adopted or pushed as the main means of telephoning, and the landlines, the proper way of telephoning, should remain as an affordable option.

The notion of "educating the nation" reflects *suomalaisuuspuhe* as it is described by Apo (1996). The Snellmanian notion of Finns needing education and *sivistys* is active in the talk about Finns' relationship to mobile phones. Also in the interview data Finnish mobile phone use was described as "*on opittu missä sitä voi käyttää, missä ei*" (people have learned where you can use it, where not), or "*ihmiset on oppinu laittamaan kiinni*" (people have learned to switch it off), or "*on vielä oppimista*" (there are still things to learn). The lack of mobile phone communication skills or bad behavior is morally evaluated in the analyzed data. Finns should know, by now, how to behave and communicate, and since they don't know, they need to be taught.

Suitable Communication Forms for Finns

In Finnishness talk, the notion of the Finns' inability to meet civilized or Central European standards is also relevant in the descriptions of Finns as mobile phone users. In the following, a woman, probably in her thirties, was attending sociologist Timo Kopomaa's public lecture at University of Helsinki in the spring of 2002. Kopomaa was invited to talk about the ways in which the mobile phone has changed the world. After Kopomaa's talk, in the question-answer period, a woman shared her experience of mobile phone behavior. Example 3:

> *mä olen ammatiltani freelance toimittaja ja mulla on periaatteessa kahdeksasta neljään työajat ja haastateltavat soittaa jatkuvasti ihan puol kymmeneen illalla. Et mä oon aatellu et niil on hirveen huono käytös tai niin ku et ku ei oo mitään niin tärkee asia et ei tarvi. Vai onks suomalaisilla hirveen huonot tavat?* (Lecture, Kopomaa, spring 2002)

> my profession is freelance journalist and in principle my working hours are from eight to four and the interviewees ongoingly call until nine thirty in the evening. I have thought that they have really bad behavior or that as they don't have anything important matters, there would be no need [to call]. Or do Finns have really bad manners?

After reporting on her observations on the change in calling habits, the increase in improper behavior, and the number of mobile phone calls she received for no good reason (*ei ole asiaa*) outside office hours, she presented a general question, "Is this typical for Finns?" Here the talk about Finn-ishness and the bad manners intersect with what is considered proper communication, that is, talking about matter-of-facts (*asiasta puhuminen*). To communicate properly, in this case, is to call *asiasta* (about a matter of importance) (Wilkins 1999) not about irrelevant matters, especially at night. Further, the woman switches from considering some individuals as not able to behave properly to talking about Finns metonymically. Not just these people who have called her, but all the Finns have bad manners when it comes to mobile phone communication.

In addition to being able to talk about significant matters at the correct time, competence in mobile phone communication is also related to the amount of talk. An interviewee described Finns and their mobile phone use with the words "*suomalainen joka ei ei puhu eikä pussaa niin, niin se pystyy sit kuitenki puhumaan [kun sillä on väline]*" (a Finn who doesn't talk or kiss, he or she is after all able to talk [when he or she has a tool]). As in the earlier data segment (Example 2), the writer wonders if the statement of Finns learning how to talk really holds true. In her letter she argues that there is now more talk but it is not meaningful and not necessarily proper, thus Finns have not learned to talk.

The collected media-data reflects years from early 1990s to 2005, and the interview data was collected in the early 2000s. From early on, Finnish mobile phone use also included writing text-messages, not just talking on the mobile phone. The writer in the following wonders if the increased amount of talk, which is a result of the wide adoption of mobile phones, is a permanent phenomenon, or something that will pass. First the reporter describes the patterns of mobile phone use of one young person, and later the reporter cites his interviewee, internet researcher Kari Hintikka. Example 4:

> *Kännykkä sai suomalaiset puhumaan ratikassakin, mutta onko sen vain ohimenevää?*
> *15-vuotias Emmi Uusi-Hakimo ei puhu kännykkään juuri koskaan. Hän pitää*
> *yhteyttä kavereihinsa tekstaamalla sekä tietokoneen keskusteluohjelmilla. Tuntuu,*
> *että puheen arvostaminen on romahtanut. Sen huomaa toimittajan työssäkin. [*
> *. . .] "vaikka suomalaisille tarjotaan yhä parempia puheaikatarjouksia, niin kir-*
> *joittaminen vaikuttaa edelleen sopivalta viestintämuodolta," hän sanoo. "Viime*
> *vuosina ovat yleistyneet esimerkiksi irtisanomiset ja parisuhteen lopettamiset*
> *tekstiviestillä.*"(NYT appendix of Helsingin Sanomat, 28/2005)

The mobile phone got the Finns to talk even in the tram, but is it just a pass-ing thing? Fifteen-year-old Emmi Uusi-Hakimo hardly ever talks on the mobile phone. She stays in contact with her friends by texting and through the computer's chat-programs. It feels as if the value of talk has collapsed. You notice that in journalist's work too. [. . .] "although Finns are offered even better offers for minutes, writing still seems to be the suitable communication form," he says. "Over the past years the lay-offs and relationship break-ups via text-messaging have become more [frequent]."

As in the data segment above (Example 4), the increased amount of talk and communication does not imply a similar increase in appropriate or proper communication (Examples 2 and 3). Writing as a form of communication is the most suitable or appropriate (*sopivin*) form for Finns. Even the most private interactions can be completed in writing. Yet, the idea includes the belief that it could be that text messaging might be a passing phase, and thus maybe Finns will remain as folk who have not learned to communicate orally. In addition, the general nature of the mobile phone is also described as inherently having those qualities that best fit Finns' needs and preferences.

Finns Relative to Others

In the late 1990s, the articles in which Finnishness and Finnish mobile phone communication were discussed compared Finland to the outside world. Usually the comparison was made to Central Europe, as in Example 5. The writers of these comparative texts commented directly or indirectly on how Finns have not "reached" the more civilized manners of the Central Europeans. Also, reporters wondered out-loud about the attention Finland was getting from the international press. In these comments, as in the next example, we can hear that what Apo (1996) has described as cultural evolutionism in *suomalaisuuspuhe*: Finland and Finns are portrayed as less civilized and less developed than other countries and cultures. The Finnish style of using mobile phones in public is considered impolite. Example 5:

> *Suomessa on suhteellisesti eniten kännyköitä maailmassa, pian 2,5 miljoonaa. Se sopii huonosti kuvaan vähäpuheisesta ja kontaktihaluttomasta ja -kyvyttömästä suomalaisesta, joka korkeintaan huutaa toiselle suomalaiselle järven toiselta rannalta. Sillä ei kai kännyköitä ihan tyhjänpanttina pidetä, vaikka ne Helsingin kesäteatterin kipeänhauskassa näytöskappaleessa hajoavatkin useammin kuin niihin soitetaan. [. . .] Suomalaistyylinen kännykkäkulttuuri ei kiehdo keskieurooppalaisia, vaikka he muuten tuppaavat hanakasti lähikontaktiin halauksin, kättelyin ja poskisuudelmin. Kännykän pirinää pidetään Keski-Euroopassa ennenkuulumattoman, epäkohteliaana kadulla, kulkuneuvoissa, kahviloissa ja muissa kohtauspaikoissa. Anteeksipyynnöt säestävät kännykänpirinää ja harvat kännykälliset säntäävät häpeillen vastaamaan vessoihin ja muihin 'piilopaikkoihin'. Mistä siis kännykkävimma on meillä peräisin? Ehkä kännykkä onkin estyneelle suomalaiselle kelpo keino päästä turvallisen etäältä sopivan lähelle toista ihmistä. Faktapohjalta, steriilisti ja vailla kosketuspintaa.* (Ilta-lehti, 070798)

Relative [to its size], Finland hosts the largest number of mobile phones in the world, soon [to be] 2.5 million. That fits badly with the image of a Finn as reticent and unwilling and incompetent to contact, who at the most shouts at the other Finn from the other side of the lake. You don't think the mobile phones are possessed for nothing, although in the painfully funny piece at the Helsinki summer theatre they break down more often than they are called to [. . .] The Finnish-style mobile phone culture doesn't interest Central-Europeans, although they otherwise are eager for close contact by hugging, hand-shaking and kissing on the cheek. Mobile phone ringing is considered in Central-Europe unforgivably impolite on the street, in transportation, in cafes and other venues for meeting. Apologies accompany a ringing mobile phone and only few mobile phone owners shamefully rush to the bathrooms and other "hiding places" to answer their phones. Where does our mobile phone enthusiasm come from? Maybe the mobile phone is for an inhibited Finn an honest way to get suitably enough close to the other person from a safe distance, for factually based reasons, in a sterile way, and without a contact surface.

Columnist Raili Nurvala above describes those concrete scenes which in Finland were part of everyday life by 1998, and which in Central Europe at that time were considered impolite or even shameful: To have a mobile phone ringing and to answer it in public transportation or on the street was considered improper. The writer explains the Finnish *kännykkävimma* (mobile phone fury) with the Finnish inhibition about being in close contact with others and with the felt safety in concentrating on sterile factual matters.

While this myth seems pervasive, some are taking issue with its prevalence. For example, a reader, who responded to Raili Nurvala's column, disagreed with Nurvala's observations on Central European and Finnish mobile phone habits and stated instead that Finns should not be ashamed of their mobile phone culture. His or her argument is based on normalcy of having mobile phones in Finland, which is interpreted here as another dimension of Finnishness talk.

Ordinary Mobile Phone and Ordinary Finns

The reader disagrees with Nurvala (Example 5) and states that extensive mobile phone use, in all environments, is now normal for the Finns. Example 6:

Raili Nurvala (7.7.) ihmetteli kolumnissaan suomalaisten kännykkäkulttuuria verrattuna keskieurooppalaisiin, jotka menevät piiloon puhumaan kännykkäänsä häpeillen. Väärin. Väitän, että Keski-Euroopassa ei vielä ole sellaista kännykkäkulttuuria, joka Suomeen on lyhyen kännykkäsesongin aikana syntynyt. Suomalaisen ei tarvitse häpeillen puhua puhelimeen, eikä se käsittääkseni häiritse enää ainakaan suurta osaa ihmisistä, vaan se on jo normaalia katukuvaa. (By Haloo in Iltalehti, 130798)

Raili Nurvala (July 7) wondered in her column how the mobile phone culture of the Finns compared to the Central Europeans, who shamefully seek to hide when talking on their mobile phone. Wrong. I claim that in Central Europe there is no such mobile phone culture that has been born in Finland during the short mobile phone season. Finns do not need to talk on their mobile phones feeling ashamed, and as far as I know it no longer disturbs at least a large number of people, but instead it is already normal in the street scene.

In addition to describing mobile phone use as normal, the pseudonymous *Haloo* above might well have noted that *kaikilla on kännykkä* (everybody has a mobile phone) or that *kännykkä on tavallinen* (mobile phone is ordinary/common), comments which were frequently heard in the research interviews. The interviewees suggested, for example, that "*suomalaisilla kaikilla on teknologia hallussa*" (all the Finns have the technology in control), or "*Suomessa kaikki puhuu koko ajan kännykkään*" (in Finland everybody speaks on the mobile phone all the time), or "*kaikilla vauvasta vaariin ja mummoon on oma matkapuhelin*" (everybody from babies to grandpas and grandmas has their own mobile phone), or "*kaikkialla Suomessa nuoret ovat sinut kännykän kanssa*" (everywhere in Finland the young people are comfortable with the mobile phone). The general shared notion seems to be that everybody throughout the whole of Finland owns a mobile phone and speaks on it all the time. The mobile phones are so *tavallisia*, so ordinary, that there is nothing to fuss about. A person with a mobile phone is *tavallinen*, and to hear and see mobile phone conversations in public is also *tavallista*.

Tavallisuus in the descriptions of owning and using the mobile phone resonates with notions on the aim of *olla tavallinen* as a feature of Finnishness (Lehtonen, Löytty, and Ruuska 2004)[3]. They discuss the ethos of being ordinary as an aspect of the strong uniform culture in Finland, and further ask, what kind of stories are told about Finnishness, and answer that one of them contains or ends with the requirement to be ordinary (Lehtonen et al. 2004, 48). To admit that one is *tavallinen* is a step taken to admitting that the speaker is taking *askeleen kansan pariin* (taking a step into or to be with the folk). As a part of Finnishness, *tavallisuus* can be proven linguistically, in interactions. To emphasize *tavallisuus* neutralizes social differences, be they economical, gender based, or age related. *Tavallisuus* is about the lack of being different, or about not being *toisenlainen* (of the other kind). The notion of *tavallinen* has a tone of unity and comfort in it, but also contains tones of irony.

Tone of Irony

In the following example the writer of the letter to the editor is again upset with the public use of the mobile phone. In particular the writer is upset with the irrelevant content of public mobile phone conversations. As in previous examples, he or she compares Finnish mobile phone use to other Europeans, and describes the ways in which Finns lack competence and skills in mobile phoning. In this example, the writer discusses mobile phone users as *tavallinen* (common or ordinary). The writer refers to Finns as *Matti* and *Maija Meikäläinen* (the Finnish version of the Americans John and Jane Doe), the ordinary, everyday, common, typical Finnish man and woman, or couple. Example 7:

Suomalaiset käyttävät muihin eurooppalaisiin verrattuna tavallista vähemmän rahaa ulkonäköönsä, vaatetukseensa ja kodin sisustukseen. Mihin sitten tämän koti-permanentatun tuulipuku-urheilusukkakansan rahat kuluvat? Vastausta ei tarvitse kaukaa hakea. Työmatkansa junalla kulkeva tietää, että kännykkälaskuihin. Eikähän siinä mitään, jos on todellista asiaa soitettavana, niin siitä vaan. Mutta jos puolen tunnin junamatkan aikana puhelin TILILIILUU TILULIILII LII jo toistakymmentä kertaa, voi sitä jo kutsua kiusanteoksi. Tai sitten kyse on huomionkipeydestä. En vain ymmärrä, että mitä pröystäilemistä on asialla, joka jo jokaisella on. Ihmettelen miten piskuinen Suomen kansa selvisikään ennen kännyköiden suomia mahdol-lisuuksia. Mahtavaa, että nyt tavallinen Matti tai Maija Meikäläinen voi tavoittaa koko maailman milloin vain ja mikä tärkeämpää, koko maailma voi tavoittaa hänet. Onhan hän tärkeä henkilö, joka on saatava kiinni heti, kun korppujauhot on kotoa ehtyneet tai kun on tiedusteltava voisiko hän ensi viikolla kotimatkallaan tuoda kioskilta puntin tulitikkuja. Tätä menoa eivät suomalaiset kohta enää osaa muulla tavalla kommunikoidakaan. Onko jo pian se päivä, kun Matti soittaa Maijalle keit-tiöön: 'No, minä täällä taas! Toisitko tänne toilettiin wc-paperia?' Tai Maija Matille: 'Vaimo tässä hei! Voisitko olla kuorsaamatta!?' Ja tähän loppuun vielä palveluidea VR:lle: Pirinävaunu. Upea laturein ja kännykkätelinein ja –hyllyin varustettu osasto meille kaikille tärkeille businessmiehille ja –naisille. Työmatkanukkujat py-sytelkööt muissa osastoissa. (By "the person who wrote this can't be reached at the moment. Please, try again," in Iltalehti, 240798)

Compared to other Europeans, Finns spend considerably less money on their appearance, clothing, and interior design. What does this *kotipermanentattu-tuulipuku-urheilusukkakansa*[4] spend its money for? One doesn't have to look far for the answer. The one who commutes to work by train knows that they spend it for mobile phone bills. That is fine if one has some real matter to call about, go ahead. But if the phone RING-RING-RING already more than ten times during a thirty-minute ride, one could call that mischief. Or it is about a need to get attention. I just don't understand what there is to brag about something that everybody already owns. I wonder how the small nation of Finland survived before the opportunities offered by the mobile phones. It is great that now the ordinary *Matti and Maija Meikäläinen* can reach the whole world whenever, and what's even more important, the whole world can reach him or her. She or he is surely an important person who needs to be reached right away when the breadcrumbs run out at home or when one needs to ask if he or she could next week bring a bundle of matches from a kiosk on his or her way home. As it goes soon Finns will not know how to communicate in any other way. Is there soon that day when *Matti* calls *Maija* in the kitchen: "Hi, it's me again! Could you bring toilet paper here to the toilet?" Or *Maija* for *Matti*: "This is your wife, hello! Could you please not snore!?" And to end this I have a service idea for the Finnish Railways: A ringing car. A great compartment with battery chargers, mobile phone racks, and mobile phone shelves for us all, important business men and women. Commuter sleepers, stay in other compartments.

What is interesting in the example above is the contradiction or agonistic dilemma in describing mobile phones and their users as ordinary and common, and simultaneously holding the belief that we are unique in the world. Also in the earlier examples, as Finnishness is described negatively, some of the texts comment also on the uniqueness of Finnishness.

Similarly to the way the writer in the Example 7 uses irony in describing Finnishness, in several interviews I conducted with Finns about their mobile phone use, I found expressions of irony.[5] In several of these interviews, the interviewees laughed at themselves for being or doing something Finnish, or for acting like everybody else. For example, an interviewee in her fifties told about a small study conducted by an individual local bus passenger. The passenger had reported his or her "findings" in a text message sent and published in a free daily newspaper. The passenger had noted that people traveling in a bus check their mobile phones every ten minutes. Piia, an interviewee, admitted while laughing that she too checks her text-messages in the bus, just like everybody else.

Lehtonen, Löytty, and Ruuska (2004, 36–41) suggest that Finnishness is commented on with a double-register: That what is considered typically Finnish is discussed with irony, with self-irony, or it is discussed with great pathos. Reading Lehtonen et al., Finns seem to feel *"lempeä ironia junttisuomalaisuutta kohtaan"* (tender or sweet irony towards the red-neck Finnishness). Also Alasuutari (1998, 169) notes the tones of irony in Finnishness talk. The use of irony in describing Finnishness and mobile phone communication is connected to the notions of *tavallisuus* or totality, and reflects a shared sense of community membership.

SUMMARY

The myth of Finnishness is recreated in several ways in the stories about Finns and their mobile phones. First of all, the plot line is that Finnish mobile phone users, just like the Finnish *kansa* (nation) on its way to independence in the nineteenth century, need education: They have not learned to communicate although the use of mobile phoning has increased dramatically. The note on the need for education is ironic when put against the apparent lack of it among Finns. The use of mobile phones for communication concentrates on talk that is not necessary, and thus not civilized. Also, mobile phone use, especially in public places, needs improvement.

Further, while describing mobile phones as a suitable form of communication for Finns, the descriptions implicitly or explicitly identify Finns as inhibited and unable to interact with others, just like the nineteenth-century agrarians were described. The most suitable way to communicate for Finns is writing, which helps to maintain the distance between the interactants.

When writing about Finnishness and their mobile phone use, whether the writings are positive or negative, it is common to make comparisons to other countries and cultures, and especially other European countries. In these comparative remarks, Finns are described as less civilized or educated, or inconsiderate mobile phone users. Also, in the writings and speech about Finnish mobile phone use, a tone of irony can be detected. Especially the tone of self-irony, reflecting playful self-criticism, is used in reporting on Finnishness.

DISCUSSION

The initial analysis was conducted on data that was produced in 1990s and in early years of the 2000s. At the time of writing this article Nokia has lost market share to Apple, and the Finnish economy, which was greatly positively impacted by the surge in mobile phone creation and manufacturing, has experienced a decline. Yet, Finns are active mobile phone, smartphone, and social media users. In 2014, every Finnish household has a mobile phone, and over 70 percent of the households have a smartphone (Official Statistics of Finland 2014).

The cultural discourse about Finnishness and the mobile phone as observed in the collected data both contests as well as recreates the over one-hundred-year-old myth of Finnishness. Finnishness talk was originally produced by educated, upper-class Finns who got to define what Finnishness and *kansa* and its qualities are. The descriptions of Finns as mobile phone users resemble *sivistyneistö*'s descriptions of the agrarian, uneducated Finns of the nineteenth century. When describing Finnishness, especially within the context of their use of technology, we find a bit more complexity than nineteenth-century mythology (on complexity, see also Alasuutari 1998).

In this analysis, I have looked at actual everyday written and spoken interactions in which Finnishness is produced by the Finns, that is, by the *kansa* (folk). Ruuska (1999, 303) suggests that as we talk about *kansa*, the speaker, whether a member of *kansa* or not, positions him or herself outside and above it, and treats *kansa* as "the other." Thus the image of Finnishness is an image given by someone better for someone of less value (Ruuska 1999, 294). When the typical Finnishness talk "takes over," the speaker describing Finnishness steps outside or above it. These steps are observable in the analyzed data as well: The speakers and writers typically do not worry and complain about "us," but about "those Finns." Ruuska (1999) also suggests that *suomalaisuuspuhe* functions as a creator of the sense of community. The other and their lifestyle remain distant and gets critically evaluated. In the case of mobile phone communication, Finnishness talk gets created and re-created in the critique that is directed towards loud, ill-placed, and too long mobile phone interactions. The interesting aspect here is that the critique considers those interactions and relations almost as one-dimensional, overlooking the fact that mobile phone conversations is a joint action between the caller and the recipient.

The production of speaking about Finnishness critically creates a sense of community from the dual roles. Every time a Finn says something ironically or critically about another Finn, the irony and critique are also about the speaker and the writer him or herself (Ruuska, 1999, 298). The ways in which Finns talk about Finnishness sound as if Finns have low self-esteem. Apo (1996) even uses the concept of self-racism, however, Ruuska (1999) or Lehtonen et al. (2004) do not agree with Apo's choice of term. Lehtonen et al. (2004, 40) point out that to be able to laugh at oneself, which they hear Finns doing, is a sign of good self-esteem, and that contradicts the image of Finnishness created in the typical Finnishness talk. Ruuska (1999, 299) instead emphasizes Finnishness talk as a recognizable way to talk about identity and to position oneself in society. For example, when the Finns criticize the supposedly uneducated and badly behaving Finns, and express a desire for their further education or cultivation, there is embedded in the talk a desire and an aim for all of the Finns to be alike (289–99). The characteristics given for Finnishness in Finnishness talk are not unique to Finns, and they are not produced in order for Finland or Finns to stand out among other nations. Instead, the critical statements reflect the shared understanding among Finns. The ways in which Finns talk about Finnishness and technology are the shared understanding among Finns, and the cultural discourse is produced within the speech community, for and with members of that community. From the perspective of ethnography of communication and cultural discourses, members of the community are those who, for example, share the interpretations for irony (for defining communication community, see, for example, Carbaugh 2007 and Philipsen 1987).

When listening to the messages about communication, social relations, and personhood in the current descriptions of Finnishness and mobile phone communication that are active in the above data, we can hear formulations of cultural premises that are summarized in the following: Communication and especially speaking "properly" on the mobile phones are an expectation. Finns as mobile phone users do not always meet the expectations, and it is suggested that they should learn to communicate. Communication is seen as something that can and should be developed or learned. Further, it is assumed that the increase in mobile phones in Finland has created more talk among Finns, and thus Finns are considered as communicating more now. Although neither the data nor the analysis allow any conclusions to be made about the changes in the everyday communication practices of Finns, the meta-communication about Finnish communication suggests that it is the quality of talk that counts as proper communication, not the quantity: People are talking more, but saying less. What is being said should be paid attention to.

The speakers and writers about Finnish mobile phone use also highlight these qualities as valued and necessary. In addition to the intellectual capacities, Finns and *kännykkäkansa* (the mobile phone nation) should be independent and able to defend themselves against, for example, economical constraints or temptations. The notion of independence is interesting in relation to the matter-of-fact description of Finnishness with less social or out-reaching qualities. Finns with their inhibitions and need for social distance have found a suitable and proper way to communicate, and this is seen as an acceptable and natural way to be. Statistics Finland no longer provides detailed information on the use of mobile phones, that is, the number of minutes called or text-messages sent. Instead, the reporting has shifted towards social media use and the use of internet via the mobile phone. In short, over 90 percent of Finns under seventy-five years of age use the internet, 70 percent uses it several times a day, and 59 percent of the respondents use internet via their mobile phones (Official Statistics of Finland 2014). For our discussion on the myth of Finnishness, the official statistics, and especially the written reports on the statistics are still offering us intriguing glimpses. Recently in a publication of Statistic Finland (Kohvakka 2013) the following description was found in the end of an article reporting on the use of social network services (for example, Facebook):

"*Yhteisöpalvelujen suosion nopea kasvu ei perustunut vain nerokkaaseen markkinointiin. Yhteisöpalveluiden runsas käyttö Suomessa kertoo siitä, että ne sopivat erityisen hyvin suomalaiseen sosiaalisuuteen. . . . Osuudet ovat kiistatta suuria: puolet suomalaisista seuraa verkkoyhteisöjään ja kolmasosa seuraa niitä päivittäin. Lisäksi suuri osa verkkoyhteisöjen jäsenistä toimii aktiivisesti yhteisössään. Kaikki suomalaiset eivät ole hiljaisia verkossakaan.*" (Hyvinvointikatsaus 2013)

Rapid increase in the use of social network services was not based only on ingenious marketing. The voluminous use of social network services in Finland tells us that they fit particularly well with Finnish sociality. . . . The shares are undoubtly large: half of the Finns follow their online social networks and one third of them does this daily. In addition, a major part of the members of the networks function actively in their networks. Not all the Finns are quiet online either.

The writer yet again, now in the context of social media use, effortlessly re-creates Finnishness talk. The writer implies that written communication, from a distance, on social media, is a match with the Finnish communication style. And in the end, he or she recognizes the silent Finn, although in a contesting way.

To be ordinary today in Finland is to have and use the mobile phone. Mobile phones and mobile phone users are ordinary. In terms of mobile phone communication, no Finn is above or beyond another Finn. To develop Lehtonen et al.'s (2004) and Apo's (1996) notions on being ordinary and totality, respectively, we could state that the mobile phone and its users are *suomalaistettu*—they have been made to be Finnish, to become Finnish or typical of Finnishness. However, since the mobile phone has also changed Finnish communication—made Finns talk more, although in a less valued way—we could ask from the data, is the Finn communicating through and on the mobile phone still a Finn?

Let me conclude by considering a comparison with other countries' mobile phone use. I recognize that I am simultaneously re-creating the comparative dimension in Finnishness talk and providing a final phase of cultural discourse analysis (Carbaugh 2007). Through comparison we find the reflection of particular cultural values. For instance, for Italians, according to Fortunati (2002, 50), the mobile phone has created new spaces for experiencing cultural values such as internal family cohesion. In the case of Israel, Lemis and Cohen (2005) studied the ways in which mobile phones in Israeli society reflect cultural values in communication. They state that Israelis have a need to communicate here and now, and this need is tied to the cultural characteristics of close familiarity, cohesive social networks, and the special security needs. The mobile phone makes this kind of communication possible and is thus valued and popular. Lemis and Cohen also found that their informants were "non-compromising and harsh in describing Israelis as hysterical, pressured, vulgar, audacious, showing off, impolite, rude, talkative, extroverted, loud, egocentric" in reporting the mobile phone related behaviors (Lemis and Cohen 2005, 197). Further, the informants stated that the inconsiderate use of the mobile phone in public was an example of the "typical" Israeli behavior, but also explainable by the unique conditions in which Israelis live, such as lack of security, feelings of pressure and anxiety, as well as caring for others and need for involvement (198).

These studies on Israeliness and Finnishness, as well as the notions on Italianness, suggest that there is in each cultural community, cultural discourse about mobile phone communication. Mobile phone communication is not a phenomenon that can exist outside its cultural context. Just the opposite, it takes place in historical and social contexts, in which there are culturally meaningful beliefs and values active about social relationships, communication, personhood, dwelling, and emotion, all which are active also in talk about and use of mobile phones.

NOTES

1. The main data analyzed to hear the stories on Finnishness and mobile phone communication were the media texts. The data includes newspaper and magazine articles of which 252 articles were published in *Helsingin Sanomat* in 1990–2004, and 122 articles in thirty-one different sources in 2001–2004. The data also includes forty-two letters to the editor, mainly published in *Helsingin Sanomat* (twenty-nine). The research interviews collected and analyzed were thematic informant and group interviews. There were in total fifty-three interviewees in forty-two interviews. Further, the data includes different kinds of public reports, studies, and other literature (etiquette books, non-fiction literature on mobile phone society, governmental reports on information society, theses and dissertations), and data on interactions such as recorded public lecturer and radio-shows, as well as some visual material (TV programs, films).

2. The source of the data excerpt is indicated in a code at the end of the original Finnish text. The code includes the date of publication or the interview and the source. For example, code (By Anne Laurila, Länsi-Suomi 260105) is read as the following: The data snipped is from Länsi-Suomi newspaper, the editor's name is Anne Laurila, and the text was published January 26, 2005.

3. Lehtonen et al.'s (2004) notions are based on Tolonen's (2002) dissertation on Finnish youth cultures in school.

4. *Kotipermanentattutuulipuku-urheilusukkakansa* is a nice example of the use, and the possibility (!), of compound words. This one word describes Finns as a nation (*kansa*) that is *kotipermanentattu* (has a homemade perm) and wears *tuulipuku-urheilusukka* combination (both parkas and sport socks). Both pieces of clothing are considered tasteless or not stylish.

5. The interviewees resided in Southern Finland, mainly in the Helsinki area. They were adults, men and women, age ranging from twenty to seventy-seven years old. The interviewees were found through friends or colleagues and their social networks, and through the social networks of other interviewees. I would describe the interviewees as "savvy social actors" (Lindlof and Taylor 2002, 177). The main body of the interviews was conducted in 2005. These interviews were recorded and transcribed, producing 225 pages of single-spaced text. The interview data also included 20 short interviews conducted at the Helsinki-Vantaa Airport asking questions of travelers about their motivations to use a mobile phone

while traveling; I was provided access to a set of interviews collected for another research project (details on these interviews are available in Toiskallio et al. 2004).

REFERENCES

Alapuro, Risto. *Suomen synty paikallisena ilmiönä 1890–1933* [The Birth of Finland as a Local Phenomenon 1890–1933]. Helsinki: Tammi, 1994.

Alasuutari, Pertti. "Älymystö ja kansakunta" [Intelligentsia and the Nation]. In *Elävänä Euroopassa. Muuttuva suomalainen identiteetti*, edited by Pertti Alasuutari and Petri Ruuska, 153–174. Tampere: Vastapaino, 1998.

Anttila, Jorma. "Käsitykset suomalaisuudesta—traditionaalisuus ja modernisuus" [Conceptions on Finnishness—Traditionalism and Modernness]. In *Mitä on suomalaisuus*, edited by Teppo Korhonen, 108–33. Helsinki: Suomen Antropologinen Seura, 1993.

Apo, Satu. "Agraarinen suomalaisuus—rasite vai resurssi?" [Agrarian Finnishness—Burden or Resource?]. In *Olkaamme siis suomalaisia. Kalevalaseuran vuosikirja 75–76*, edited by Pekka Laaksonen and Sirkka-Liisa Mettomäki, 176–184. Helsinki: Suomalaisen Kirjallisuuden Seura, 1996.

Apo, Satu. "Suomalaisuuden stigmatisoinnin traditio" [Tradition of Stigmatizing Finnishness]. In *Elävänä Euroopassa. Muuttuva suomalainen identiteetti*, edited by Pertti Alasuutari and Petri Ruuska, 83–128. Tampere: Vastapaino, 1998.

Berry, Michael. "If You Run Away from a Bear You Will Run into a Wolf: Finnish Responses to Joanna Kramer's Identity Crisis." In *Texts and Identities: Proceedings of the Third Kentucky Conference on Narratives, 1994*, edited by Joachim Knuf, 32–48. University of Kentucky: Lexington, 1995.

Carbaugh, Donal. *Situating Selves. The Communication of Social Identities in American Scenes*. Albany, NY: SUNY Press, 1996.

Carbaugh, Donal. "Cultural Discourse Analysis: Communication Practices and Intercultural Encounters." *Journal of Intercultural Communication Research* 36, no. 3 (2007): 167–182. doi:10.1080/17475750701737090.

Fortunati, Leopoldina. "Italy: Stereotypes, True and False." In *Perpetual Contact: Mobile Communication, Private Talk, Public Performance*, edited by James Katz and Mark Aakhus, 42–62. Cambridge, MA: Cambridge University Press, 2002.

Gordon, Tuula, Katri Komulainen, and Kirsti Lempiäinen. "*Suomineitonen hei! Kansallisuuden sukupuoli*" [Greetings Young Girl Finland! Gender of Nationality]. Tampere: Vastapaino, 2002.

Häikiö, Martti. "Kännylcät ja suamalainen kansanluonne" [Mobile Phones and the Finnish Folk Psychology]. *Kanava* 9 (1998): 571–72.

Kohvakka, Rauli. "Yhteisöpalvelut istuvat suomalaiseen sosiaalisuuteen" [Social Network Services Fit into Finnish Sociality]. *Hyvintointikatsaus*, 2 (2013): 58–62. Helsinki, Finland: Tilastokeskus.

Lehtonen, Mikko, Olli Löytty, and Petri Ruuska. *Suomi toisin sanoen* [Putting Finland in Other Words]. Tampere: Vastapaino, 2004.

Lemis, Dafna, and Akiba A. Cohen. "Tell Me about Your Mobile and I'll Tell You Who You Are: Israelis Talk about Themselves." In *Mobile Communication:*

Re-Negotiation of the Social Sphere, edited by Richard Ling and Per E. Pedersen, 187–202. London: Springer, 2005.

Leskinen, Tatu. "Kansa, kansallisuus ja sivistys" [Nation, Nationality and Education]. *Jargonia* 6, no. 3 (2005). Accessed January 20, 2015. https://jyx.jyu.fi/dspace/handle/123456789/20064.

Lindlof, Thomas R., and Brian C. Taylor. *Qualitative communication research methods*, 2nd edition. Thousand Oaks, CA: Sage, 2002.

Mäenpää, Pasi. "Tietoyhteiskunta ja uusi suomalaisuus" [Information Society and New Finnishness]. In *Suomen kulttuurihistoria, osa 4. Koti, kylä, kaupunki*, edited by Kirsi Saarikangas and Minna Sarantola-Weiss, 517–8. Helsinki: Tammi, 2004.

Nieminen, Hannu. "Medioituminen ja suomalaisen viestintämaiseman muutos" [Mediazation and the Change in Finnish Communication Scene]. In *Uusi media ja arkielämä. Kirjoituksia uuden ajan kulttuurista*, Publications A, No. 41, edited by Hannu Nieminen, Petri Saarikoski, and Jaakko Suominen, 18–43. Turku: University of Turku, School of Art Studies, 2000.

Official Statistics of Finland (OSF). "Use of Information and Communications Technology by Individuals" [e-publication]. Helsinki: Statistics Finland, 2014. Accessed January 15, 2015. http://www/stat.fi/til/sutivi/2014/sutivi_2014_2014-11-06_tie_001_en.html.

Peltonen, Matti. "Omakuvamme murroskohdat. Maisema ja kieli suomalaisuuskäsityksen perusaineksina" [Turning-Points of Our Self-Image. Landscape and Language as the Basic Ingredients of the Conception of Finnishness]. In *Elävänä Euroopassa. Muuttuva suomalainen identiteetti*, edited by Pertti Alasuutari and Petri Ruuska, 19–40. Tampere: Vastapaino, 1998.

Philipsen, Gerry. "Speaking 'Like a Man' in Teamsterville: Culture Patterns of Role Enactment in an Urban Neighborhood." *Quarterly Journal of Speech* 61, no. 1 (1975): 13–23.

Philipsen, Gerry. "The Prospect for Cultural Communication." In *Communication Theory: Eastern and Western Perspectives*, edited by Lawrence D. Kincaid, 245–54. Orlando, FL: Academic Press, 1987.

Ruuska, Petri. "Negatiivinen suomalaisuuspuhe yhteisön rakentajana" [Negative Finnishness Talk Building a Community]. *Sosiologia* 36, no. 4 (1999): 293–305.

Ruuska, Petri. *Kuviteltu Suomi. Globalisaation, nationalismin ja suomalaisuuden punos julkisissa sanoissa 1980–90-luvuilla* [Imagined Finland: Interweaving Globalization, Nationalism and Finnishness in Public Discourse in the 1980s and 1990s]. Acta Electronica Universitatis Tamperensis No. 156. Tampere: Tampereen yliopistopaino, 2002.

Toiskallio, Kalle, Sakari Tamminen, Heini Korpilahti, Salla Hari, and Marko Nieminen. *Mobiilit käyttökontekstit—Mobix. Loppuraportti* [Mobile Contexts of Use—Mobix. Final Report]. Software Business and Engineering Institute. Helsinki University of Technology, 2004.

Varpio, Yrjö. *Pohjantähden maa. Johdatusta Suomen kirjallisuuteen ja kulttuuriin* [Country of the North Star. Introduction to Finnish Literature and Culture]. Tampere: Tampere University Press, 1999.

Vehviläinen, Marja. "Teknologinen nationalismi" [Technological Nationalism]. In *Suomineitonen hei! Kansallisuuden sukupuoli*, edited by Tuula Gordon, Katri Komulainen, and Kirsti Lempiäinen, 211–29. Tampere: Vastapaino, 2002.

Wieder, Lawrence, and Steven Pratt. "On Being a Recognizable Indian among Indians." In *Cultural Communication and Intercultural Contact*, edited by Donal Carbaugh, 45–64. Hillsdale, NJ: Lawrence Erlbaum, 1990.

Wiio, Osmo A. "Matkapuhelimen voittokulku" [Triumphal March of the Mobile Phone]. Review of *Alkuräjähdys—Radiolinja Suomen GSM-matkapuhelintoiminta 1988–1998*, by Martti Häikiö. *Kanava*, no. 8 (1998): 513–5.

Wilkins, Richard. "'Asia' (Matter-of-Fact) Communication: Finnish Cultural Terms for Talk in Education Scenes." PhD dissertation, University of Massachusetts at Amherst, 1999.

Wodak, Ruth, Rudolf de Cillia, Martin Reisigl, and Karin Liebhart. *The Discursive Construction of National Identity*. Edinburgh: Edinburgh University Press, 2009.

SEVEN

Intentional Design

Using Iterative Modification to Enhance Online Learning for Professional Cohorts

Lauren Mackenzie and Megan R. Wallace

INTRODUCTION

The following words, drawn from Flyvbjerg's (2001, 166) *Making Social Science Matter*, speak to those committed to making *course design* matter: "we must take up problems that matter to the local, national, and global communities in which we live, and we must do it in ways that matter . . . [and] we must effectively communicate the results of our research to fellow citizens." This chapter emphasizes the importance of incorporating the lived experiences of professional students into the instructional design process. In an increasingly online educational world, this chapter contributes to the ongoing conversation about intentional design by putting forth a formula for course development. In doing so, the authors examine the diverse cultural practices of military students in an online intercultural communication course offered by the Community College of the Air Force and drawn from a cultural community of over 2,000 military students who have written about their cross-cultural experiences in the course wiki.[1]

The authors have been documenting the ongoing progress of the "Introduction to Cross-Cultural Communication" (CCC) course since its inception in 2009 and pilot in 2011. When the course first opened in 2011, it was completely self-paced and did not include opportunities for students to contribute to the course content. Now, four years later, there is an active course wiki—which has led to the creation of a variety of Situational Judgment Tests (SJTs) based on students' intercultural experiences. This iterative course modification process is the connection to instructional design and the main focus of this chapter. Previously, publications have been devoted to the challenges and opportunities associated with the asynchronous nature of the CCC course (Mackenzie and Wallace 2012), the function, utilization, and consequences of the course wiki (Mackenzie and Wallace 2014), as well as the techniques used in the course to increase student retention (Mackenzie, Fogarty, and Khachadoorian 2013).

The current chapter builds on this work by situating it in a Cultural Discourse Analysis (CuDA) framework with a focus on the Situational Judgment Test as a teaching and learning tool that is inherent in the course design. In this particular course, the educational design process requires knowledge of military-specific contexts to meet the needs of the institution as well as the students. Consequently, the SJT is an appropriate teaching and learning tool that makes communication "not only its primary data but moreover, its primary theoretical concern" (Carbaugh 2007, 167). By analyzing the communicative practices associated with constructing SJTs in this particular course, the authors have devised a culture-specific, military-appropriate, and communication skill-centered formula for the online military culture classroom.

An increasingly diverse workforce has led more professions than ever before to address cross-cultural competence in their training, education, and research programs. The disciplines of medicine (for example, Jeffreys 2008; Crosson et al. 2004; Crandall et al. 2003), law (for example, Bryant 2001), social work (for example, Teasley 2005), business (Cox 1994; Miller 2006), and education (for example, Barrera and Corso 2002) have all begun to integrate, to some extent, the idea that competent cross-cultural communication is an essential component of professional competence. The CCC course described in this chapter addresses this concern within the military, a profession in which the outcome of failed cross-cultural communication may have fatal consequences (see Nelson 2007). Because being cross-culturally competent can mean the difference between success and failure in a variety of careers, it is important that the training for these professionals be designed to have the best chance of success. Consequently, this chapter's focus is on designing programs of instruction for culturally distinct cohorts that are both effective and appropriate—utilizing a course for American Airmen as a case study. The literature regarding effective and appropriate professional instruction will be reviewed forthwith to properly situate the present discussion within extant scholarship.

REVIEWING THE LITERATURE: ONLINE
LEARNING (OLL) CONSTRAINTS AND ENABLERS

Instructional Design: Teaching a Professional Cohort

Web-based course systems have been praised due to the ability to ease the delivery of professional development (Artino 2008; Branzburg and Kennedy 2001; Fenton and Watkins 2007; Holzer 2004; Sandars and Langlois 2005; Santovec 2004; Weingardt, Cucciare, Bellotti, and Lai 2009) while maintaining quality and effective instruction (Artino 2008; Moneta and Kekkonen-Moneta 2007). By presenting increased asynchronous, or self-paced options, instructors can offer more convenient opportunities to accommodate professional student schedules (American Society for Training and Development 2005; Artino 2008; Hew, Cheung, and Ng 2010; Hew and Cheung 2012; Kelly 2005; May, Acquaviva, Dorfman, and Posey 2009). Asynchronous courses inherently increase the need for students to be motivated and disciplined to succeed (Lorenzetti 2004; Murphy, Rodríguez-Manzanares, and Barbour 2011; Short 2000); however the ability to reach a greater number of students typically hindered by distance or scheduling more than made up for these minor setbacks. The advantages are even more evident when working with those employed in the armed services who often have challenging schedules and may be working anywhere in the world.

Instructional design for online learning (OLL) shares similarities with the process of creating face-to-face instruction; however significant differences are noted. In both cases, it's considered to be an art as well as a science (Botturi 2006; Kenny, Zhang, Schwier, and Campbell 2005) and should be both an iterative and organic process (Gustafson and Branch 2002). Due to the increased complexity of OLL, a team approach to designing online courses is the best practice (Holsombach-Ebner 2013; Restauri 2004). The most successful OLL courses simulate the features of face-to-face classes that work well (Hew and Cheung 2012; Manning 2007; Rempel and McMillen 2008) incorporating the essential components of interaction: the instructor, fellow students, and the material (Licona and Gurung 2011; Swan 2003). This can be accomplished through the incorporation of video, audio, discussions, wikis, virtual classrooms, depending on the available features of the academic content delivery platform utilized.

Course Management Systems (CMS)

As for many instructional designers, the course or content delivery management system (CMS) was predetermined by the institution (Holsombach-Ebner 2013; Rempel and McMillen 2008). The most commonly employed CMS platform is Blackboard (Bb) (Bradford, Porciello, Balkon, and Backus 2007; NCDAE 2006; Snow and Sampson 2010), and this is certainly the case for the Air Force. The enabling features of Blackboard include a "classroom" feel, discussion boards, wikis, widely available tech support both within the university and from Blackboard itself, and its familiarity to many students (Holsombach-Ebner 2013; Licona and Gurung 2011; Rempel and McMillan 2008). The Bb interface offers students a designated shared space in the digital classroom (Brunk-Chavez and Miller 2007) activating explicit group identities (Licona and Gurung 2011). Most importantly, it helps the students engage in a community of practice (Lave and Wenger 1991) which the CCC course is attempting to create: Airmen with the ability to employ cross-cultural communication skills professionally.

For OLL designers, it is all about maximizing the features of the CMS platform to improve the effectiveness and appropriateness of the instruction (Allen and Seaman 2013). In OLL, it is inherently more difficult to ensure interaction goals are met; therefore including tools like discussions, virtual classrooms, and wikis to overcome the challenges of interaction in online courses is essential (Cleaver 2008). For asynchronous courses, the most heavily employed tool is the online asynchronous discussion (OAD) (Beckett, Amaro-Jiménez, and Beckett 2010; Holsombach-Ebner 2013). Levin, He, and Robbins (2006) defined OADs as enacted social construction theories in the virtual environment, and they involve three S's: Self interaction, Subject interaction and Social interaction (Licona and Gurung 2011). OADs have become part of standard accepted OLL design too (Blackboard 2012). Because OLL offers so many advantages in relation to the disadvantages, overcoming the perceived lack of interaction in virtual classrooms seems a surmountable task.

Meeting Student Interaction Needs

The OADs offer myriad benefits to OLL including increased student engagement to enhance learning (Blankenship 2011; Dringus and Ellis 2009; Lin 2008; Rempel and McMillen 2008; Roberts 2002; Rowley and O'Dea 2009) with improvements in student outcomes (Lin 2008; Larson and Sun 2009; Xia, Fielder, and Siragusa 2013; Zha and Ottendorfer 2011). Additionally, OADs have been found to successfully emulate the community feel of a traditional classroom (Beckett, Amaro-Jiménez, and Beckett 2010; Bryce 2014; Lord and Lomica 2004; Rempel and McMillan 2008; Xia, Fielder, and Siragusa 2013). OADs allow students to share and discuss specific knowledge, such as experiences and terminology (Beckett, Amaro-Jiménez, and Beckett 2010; Lee and Tsai 2011; Rempel and McMillan 2008), proving valuable for professional cohort socialization (Beckett, Amaro-Jiménez, and Beckett 2010).

One alternate format for OADs offered is the wiki. Wiki is defined as loosely-structured, collaboratively edited web-linked content on a particular subject (Beldarrain 2006). Several features that distinguish the wiki in OLL from a traditional discussion board include its nonlinear structure and designation as a collective body of knowledge rather than a threaded conversation (Mackenzie and Wallace 2014). Wikis are particularly suited for institutional learning due to its creation of an enduring product (Lackey 2007; Murphy 2004; Rivait 2014) that can be useful for both students and the organization to define shared professional knowledge (Szabady, Fodrey, and Del Russo 2014). Because synchronous chat just isn't feasible with the schedules of today's online students (Manning 2007), features like wikis allow for aggregated student responses to inform the course outcome in a similar fashion to face-to-face class discussions. This encourages online students to help build the shared body of knowledge instead of remaining static learners (Cleaver 2008). These student narratives contain copious amounts of rich qualitative data and although the present chapter is not the first to publish about examining student wiki contributions (see Hara, Bonk, and Angeli 2000; and Picciano 2002) this endeavor is novel in its treatment of OAD content as cultural discourse.

For the CCC course, the familiar "wiki" moniker and format are used to emphasize the shared ownership of their collaboratively generated knowledge. Students are not asked to respond to or discuss with each other as each student's individual narrative will become part of the group narrative for each topic. Bb discussion boards are complex and layered (Kuhlenschmidt 2009), but the wiki is designed to only be one subject deep (Center for Instructional & Learning Technologies 2010) and thus can search for a unified description of the students' lived experience. Each wiki prompt acts as an open-ended interview question. The group collectively determines the content of the wiki (Kuhlenschmidt 2009), which then can evolve into a shared story as often occurs in focus groups. However, there are two significant shortcomings of the wiki design within the Blackboard CMS. The first is that students are locked out and unable to add to or edit the wikis for 2 minutes if another student is editing (Blackboard, Inc. 2013) which may create serious conflicts due to the large class size and demanding schedules (for example, deployments, twenty-four-hour shifts). It is essential to professional student cohorts that the entire course be available to them when they need it. Another shortcoming of the Blackboard wiki set-up is that grades are not individually assigned (Blackboard, Inc. 2013). If credit is given to wiki participation, instructors need to be certain of a student's contribution to assign credit, and this process is time consuming within the CMS because it requires the instructor to manually edit individual grades. Due to these considerations, the course designers elected to set up the wiki to record additions by student name, which within the Blackboard CMS can be quite similar in appearance to a traditional discussion board when organized in this manner. Because Blackboard wiki functionality has evolved so much over the four years of this course, students and staff may not be familiar with using the technology, and increased support needs should be taken into account when deciding to employ a wiki, especially for courses with large enrollment numbers.

Situational Judgment Tests

Given that CCC enrolls hundreds of students each semester, it became necessary to find creative ways to bring course content to life while assessing student learning. The use of SJTs has proven to be an effective assessment tool and means for applying communication-centered concepts and skills via culture-specific, military-appropriate scenarios. Before sample, validated SJTs are discussed; however, it is important to provide a brief overview of the SJT literature.

SJTs are often described as selection procedures involving job-related situations that are presented with multiple-choice response options (Krumm, Lievens, Hüffmeier, Lipnevich, Bendels, and Hertel 2014). SJTs assess the "ability to perceive and interpret social dynamics in such a way that facilitates judgments regarding the timing and appropriateness of contextual behaviors" (Christian, Edwards, and Bradley 2010, 92). Dozens of empirical studies devoted to SJTs have been published since 1990 (Campion, Ployhart, and MacKenzie 2014), most of which found SJTs to be effective measures of leadership competencies. Of particular interest to the current project is the research devoted to intercultural SJTs since intercultural interactions are not only complex and challenging, but also prone to misjudgment (Ang and Van Dyne 2008; Earley and Ang 2003). This is due, in part, to the fact that others don't often explicitly communicate their expectations of appropriate behavior (Molinsky 2013). The authors' research devoted to the CCC course is one means of addressing this complexity and offers culture-general concepts and skills followed by culture-specific SJTs for application practice and assessment.

Although there are a variety of ways to measure intercultural competence, the SJT offers a more context-specific means for capturing the complexity of intercultural interaction than the self-report instruments that are most commonly used to measure intercultural competence (Leung, Ang, and Tan 2014). As noted by Rockstuhl, Ang, Ng, Lievens, and VanDyne (2014, 14), the intercultural SJT provides "an alternative performance-based assessment tool that has good predictive validity" for measuring intercultural competence. A study supporting this claim was conducted by Rockstuhl, Ng, and Ang (2013) demonstrating that performance on intercultural SJTs has predicted supervisor-rated task performance of Filipino professionals.

Regardless of the professional cohort or culture participating in the SJT, after reading a job-related scenario, the user is typically asked a question such as (McDaniel, Whetzel, and Nguyen 2006):

1. "What would you do next?"
2. "What would you be most likely and least likely to do?"
3. "What is the best response among all options?"
4. "What would most likely occur next in this situation as a result of your decision?"

The SJTs found in the authors' cross-cultural communication course utilize variations of all four of these questions to assess student learning. In line with traditional competency measures which suggest a grounded theory approach to understanding specific skills required in particular jobs (Spencer and Spencer 1993), the authors created the SJTs by utilizing students' wiki contributions to align military experiences with the course content. Analysis after three iterations of the course provided construct validity for the SJTs—which positively correlated them with the overall course average as well as post-course intercultural knowledge and flexibility measures.[2]

Working with the assumption that SJTs are a valid and effective means for assessing cross-cultural knowledge and skills, the current chapter aims to contribute to the existing work by offering an intentional design formula for cross-cultural communication SJTs in military distance education. Campion et al.'s (2014) *The State of Research on Situational Judgment Tests: A Content Analysis and Directions for Future Research* suggests that future research should "more strongly incorporate theory into the design and development of SJTs" (303). Further, the authors posit that an "interactionist" perspective might advance SJT research by reminding test-makers that, "to study the person in the situation, one needs to study how the person *interprets* the situation" (304).

This is where the current chapter contributes to that call—investigating how military students (as "users" of a large CCC course) bring to life the communication skills introduced in the course by contributing reflections about their military cross-cultural experiences. These initial reflections are then converted into culture-specific scenarios used to illustrate the course concepts and assess students' ability to apply them in military-appropriate situations.[3] After the pilot of the CCC course was complete and the authors read the end-of-course student evaluations, it became necessary to include an opportunity for students to contribute to the course content. In fact, the very nature of drawing from the lived experiences of the students is what makes the course complete. Although the authors sought to bring CCC to life on students' screens with a variety of media and interactive software, it was ultimately the students' intercultural experiences that brought the content to life for their colleagues by giving the communication concepts and skills military relevancy.

Cultured Organizations and Engaged Scholarship

Organizational culture cannot be overlooked when considering how to communicate appropriately with a professional cohort. An organization's culture consists of "webs of meaning" created through communication (Pacanowsky and O'Donnell-Trujillo 1983) and defined by interaction (Bormann 1983). This is extremely evident in the evolution of modern military culture (Katzenstein and Reppy 1999). Each new member comes to the service with their own cultural behaviors, beliefs, and identities (Varvel 2000); however, interaction between diverse Airmen along with Air Force doctrine serves to help define the culture (Poyner 2007). The military relies on this collective identity and promotes intensive enculturation during basic training (Katzenstein and Reppy 1999). Practicums and internships serve a similar purpose in many civilian careers.

When the uniqueness of organizational cultures and the necessity for enculturation are taken into account, the usefulness of engaged scholarship is evident. Engaged scholarship is defined by Van De Ven (2007) as a variety of participative academia in which scholars perform research while fully immersed within an organization or discipline. This *in situ* perspective allows for contact with the full gamut of stakeholders within the cultural context and exposure to the most relevant issues facing these professionals. The advantage is such that "by involving others and leveraging their different kinds of knowledge, engaged scholarship can produce knowledge that is more *penetrating* and *insightful* than when scholars work on the problems alone" (Van De Ven 2007, 9; emphasis added).

Contextualizing academic material within relevant practical applications can help professional students engage in critical thinking (Chandler 2005). To be entirely appropriate to the target professional group, culture must be taken into account as "a workplace requires practitioners to seek fundamentally different ways of responding to their contexts and exigencies" (Alred 2006, 81). So, how does an instructional designer ensure that a particular groups' "contexts and exigencies" are addressed appropriately and in their vocabulary? Ethnography of communication is a prominent way to study culture and communication, and student discourse is the essential human data.

Methodological Considerations

The CCC course itself—as well as the methodology used to inform its iterative modification—is indebted to the ethnography of communication (EC). This section of the chapter will review how EC has informed the CCC course and include the content and context of CCC itself. The remainder of the chapter will be devoted to an intentional design formula for constructing military culture SJTs, and conclude by connecting intentional design with engaged scholarship.

The course (and the research devoted to it) uses an EC orientation toward the study of cross-cultural communication, viewing communication generally as "the primary social process" (Carbaugh 1990, 18) and specifically as: "a sociocultural system of coordinated action and meaning, that is, an interactional system that is individually applied, socially negotiated, symbolically constituted and culturally distinct" (Carbaugh and Hastings 1992, 159). This approach to studying cross-cultural communication makes the presumptive claim that language *use* cannot be separated or even understood apart from the scenes in which it occurs, and that specific emphasis must be placed on the study of communication practice itself. In particular, EC has built knowledge about communication by presuming the following: that "everywhere there is communication, a system is at work; that everywhere there is a communication system, there is cultural meaning and social organizations and thus, that the communication system is at least partly constitutive of socio-cultural life" (Philipsen 1992, 7–16). To maximize the effectiveness and appropriateness, instructional material for professional cohorts must emphasize the context of communication and be situated within the organization's socio-cultural environment. Thus, EC is conceived of not only as a research tool, but also as a means for students to understand culture and for educators to provide culturally appropriate ways to teach.

In addition to its focus on locally distinctive practices of communication, EC also has informed the CuDA methodology which offers procedures for analysis of communication practices as formative of social life (Carbaugh 2008). This approach to communication and culture is particularly useful for military personnel who need to understand culture in both general and specific ways and who will be experiencing communication in a particular context, but in a wide variety of cultural settings. Although culture is researched and taught by military scholars in a variety of disciplines (mainly political science, international relations, sociology, and anthropology), as noted by Carbaugh (1988b, 40):

> The culture concept is used best in our empirical studies when it describes communication patterns of action and meaning that are deeply felt, commonly intelligible, and widely accessible, and when it explores situated

contexts of use through conceptual frames, treats cultural terms as focal
concerns, and exploits the benefits of comparative study.

The authors aimed to answer Carbaugh's call, particularly the focus on
situated contexts in a professional setting. Further, this study is situated
within an ongoing program of CuDA work—which systematically explores language "in use" and treats communication as cultural discourse.
The authors drew from studies that have featured prominently both the
"context" and the participants' membership and identities as shaped
through dialogue. For example, Carbaugh, Nuciforo, Saito, and Shin's
(2011) analysis of the distinctive features of "dialogue" in Japanese,
Korean, and Russian discourse; Witteborn's (2011) examination of intercultural dialogue in a virtual forum via the use of "identity terms" and
"truth talk," among others; Milburn's (2009) cultural discourse analysis
of "membership" and "community context" in non-profit organizations;
as well as Miller and Rudnick's (2008) work devoted to the *Security Needs
Assessment Protocol* for the United Nations in which they research local,
cultural discourses in order to develop effective and appropriate strategies for working within a community.

A summary of the course's objectives, concepts, and skills will now be
provided to illustrate how this EC approach to culture was infused into
the design and development of the CCC course.

COMMUNICATION CONTENT IN
A MILITARY CONTEXT: AN OVERVIEW OF
THE CROSS-CULTURAL COMMUNICATION COURSE

To best serve its student population, the CCC course is situated squarely
where professional development meets academic instruction. Built on
a solid base of quality scholarship from the discipline of communication, the course focuses on applying the field to the military profession.
CCC is offered at no cost and available to all enlisted Airmen by the Air
Force Culture and Language Center (AFCLC) via Blackboard.[4] Under the
accreditation purview of both Air University and the Community College of the Air Force (CCAF) since 2011, the course fulfills three general
elective or social science credits and contains twelve lessons. Beyond the
course credit assignment, the course's focus on communication as part
of relationship-building also fulfills professional development needs that
align with the twenty-first-century Air Force readiness goals. As has been
stated in a variety of military publications, military power is no longer
just about firepower, it about the power to build relationships (Ben-
Yoav Nobel, Wortinger, and Hannah 2007). The military recognizes that
cross-culturally competent communication is one of the keys to building

successful relationships, and the authors of this chapter and of the CCC course are committed to ensuring that communication remains at the center of this effort.

As such, there are three objectives of CCC which collectively facilitate the development of student service members' cross-cultural communication competence and can be categorized into knowledge, skills, and abilities. First, the course familiarizes students with the leading concepts, theories, and scholars of cross-cultural communication and seeks to instill in students a sense of the importance of competent cross-cultural communication in both personal and professional settings. Second, it introduces students to the skills that comprise cross-cultural competence (3C). Lastly, and most importantly, it enables students to apply cross-cultural communication skills in a variety of Air Force contexts. It is the field-specific SJT applications that transform this course from a cross-cultural communication course taken by students in the military to a cross-cultural communication course for military students.

Cross-cultural communication is a new discipline in military scholarship, as it was once in the fields of education, counseling, and medicine. Intentional design was essential due to the necessity to make the case for communication as an indispensable professional competency in matters of national defense (see Mackenzie 2014, and Greene-Sands and Greene-Sands 2014, for arguments devoted to institutionalizing intercultural communication into Professional Military Education). The success of the classes' instructional design methods highlights its potential to inform cross-cultural communication training and education in a wide variety of disciplines. Although created for Air Force members, the structure and profession-specific SJT formula are applicable to cross-cultural competence training and development for any professional cohort.

The first course objective relating to knowledge is achieved through eleven lessons of cross-cultural communication content (also described in Mackenzie and Wallace 2012) focused on the most up-to-date scholarship and military applicability. The ability to situate communication within the context of the military community, as well as within greater academia, is a consequence of the position of engaged scholarship afforded by the Air Force Culture and Language Center. The course content never loses focus of the applicability of every concept to the military profession, its singular purpose being to build a comprehensive toolkit for members' intercultural interaction. The essential concepts are divided into eleven content lessons devoted to:

- Culture-General vs. Culture-Specific
- Cross-Cultural Communication is about Interpreting Cultural Messages

- Narrative Functions to Communicate Identities Across Cultures
- Communication Approaches to Culture
- Cross-Cultural Communication Competence
- Managing Paralanguage Use and Perception
- Decoding Nonverbal Messages
- Active Listening
- Identifying and Adapting to Different Communication Styles
- Employing Effective Interaction Skills
- Building Relationships and Managing Conflict

The second course objective ("introduces students to the skills that comprise cross-cultural competence") is achieved by a variety of visual enhancements. As previously described, OLL brings unique challenges and opportunities to the communication classroom. To address the interactional shortcomings, each lesson includes an introductory video by the professor lasting approximately five minutes, followed by a movie clip application, twenty pages (on average) of course content per lesson, an average of two readings (one to two hours), interactive knowledge checks, and scenario-based exercises (SJTs). The readings are selected intentionally to be both military-relevant and academically-based to balance academic rigor with professional applications. Instruction is typically followed by video illustrations and case studies of communication successes and failures. The course is filled with examples of successful cross-cultural communication as well as failed attempts and the ways in which they can impact both personal and professional relationships. An emphasis on the skills associated with competent cross-cultural communication reinforces the importance of mission-effective *and* culturally appropriate communication for military students.

The final course objective, ("enabling students to apply cross-cultural communication skills in a variety of Air Force contexts") is achieved through the instructional design strategy. Given the objectives of the course and the online delivery of the content, the instructional strategy must include methods of engaging students in the content, and with each other, in a way that meaningfully contributes to the acquisition and retention of communication skills. The study of communication is inherently contextual, and is about language *use* (not language production). Consequently, through scenarios, case studies, and video illustrations students are exposed to successes and failures in cultural message interpretation. They are also provided opportunities to engage in active listening, perspective-taking, and interpreting cultural messages in context through the use of scenario-based learning followed by Situational Judgment Tests. The majority of the Situational Judgment Tests have a military operations focus and serve to remind students of the connection between the quality of their communication and the quality of their relationships. The profession-specific SJTs naturally follow this confluence of academic knowledge and practical skills. Finally, the twelfth lesson provides students with an opportunity to apply what they've learned via a live-actor, film-based, cross-cultural immersion scenario.

INTENTIONAL DESIGN: AN SJT FORMULA
FOR PROFESSIONAL ONLINE CULTURE COURSES

The CCC pilot in 2011 included several non-military, non-communication-centered SJTs borrowed from international business articles. By 2014, however, all lessons in the course include original SJTs informed by students' intercultural experiences and written by the authors. Communication in skills practice is the theoretical concern and focus of inquiry for the SJTs and drove the authors to develop an intentional design formula for each lesson of the course. This process is discussed in the section to follow.

The CCC SJTs have been gleaned from previous student wiki contributions about their deployments and overseas assignments to introduce cross-cultural communication skills and assess student understanding. Military students' experiences have become teaching tools for themselves and their peers, utilizing an intentionally diverse sample of locations for the culture-specific scenarios (examples of countries included are: Singapore, Italy, the Bahamas, S. Korea, and Canada, among others). Generally speaking, these scenarios assess a student's ability to apply communication concepts and theories in culturally complex circumstances, and thus require higher-order thinking and decision-making on the part of the student. Scenarios are designed to (1) be consistent with the types of situations military students have faced while deployed or stationed abroad, (2) utilize the language and code of the military culture, and (3) provide immediate feedback to the student about how optimal each choice is compared to the other options presented. Thus, the testing experience itself is a formative learning tool, allowing students to experience simulated consequences of their cultural choices.

The question then remains: how does an instructor move from reviewing wiki contributions to SJT construction? This question can be answered by drawing from grounded theory and CuDA. The wiki "experience" by the authors and CCC professors is inherently ethnographic—that is, the professors are *in situ* with our students in the only way you can be in an online, self-paced course. The Blackboard wiki format assisted in aggregating student responses by overarching themes within the context of each lesson. The designers utilized a process of idiographic coding with a focus on individual units of meaning and nomothetic coding with a broad focus to look for overarching and more abstract themes (Grbich 2007).

One means for turning an online, self-paced course (created initially to introduce CCC academic concepts and skills) into a professional student-centered learning experience is to solicit students' contributions that connect to the course content. It was the researcher-instructors' way of reminding military students that this isn't just any online course—but instead a course infused with previous students' intercultural experiences to bring the content to life and ensure relevance to the military cohort.

A variety of wiki prompts are used in all eleven content-based lessons of the course and ask questions such as:

Lesson 5: Provide an example you've witnessed of effective but not appropriate or appropriate but not effective communication.
Lesson 10: Think about the most difficult conversation you've had with a person from a different culture. Now that you've learned about the importance of impression management (that is, self-monitoring, emotion regulation, and perception checking), describe what you could have done differently to improve the outcome of the conversation.

Thus, as the authors began to review the hundreds of wiki contributions at the end of the course, the following questions were asked in order to set apart the contributions that would be transformed into an SJT for the next iteration of the course.

1. Is there a new culture or Air Base that many CCAF students have been assigned to that has not yet been mentioned? Many culture-specific stories and examples found in the course are Middle-Eastern (since that is where the majority of military operations have occurred since the course began), and several students requested more diversity in the cultures represented.
2. Is there a specific communication practice related to the lesson that has not yet been elaborated on in the course? Although the course introduces a wide range of communication practices and speech acts, some resonate with military students more than others. The wiki contributions give the professors an opportunity to find patterns that should be represented in the course.
3. Is there a different relationship written about that could put the communication concept/skill into a new context? For instance, the majority of the scenarios described in the anecdotes/examples throughout the course were adapted from general academic or business literature and were devoted to roommates, significant others, and supervisors/Commanders. The students' wiki contributions

gave the authors an opportunity to learn about other relationships that are of significance and/or problematic to military students.

4. Is there a military symbol, ritual, or object that could lend the lesson content more military legitimacy? Since the CCC Professors are/have been married to military personnel and lived on many military installations but not served themselves, this question helps ensure that the course examples and SJTs are up-to-date and consistent with the lived experience of an enlisted Airman. Using military symbols, rituals, and objects in the SJTs that are gleaned from previous students' wiki contributions ensures military appropriateness in the culture-specific examples provided throughout the course.

These questions reflect the authors' intent to heed Carbaugh's call to "explore situated contexts of use through conceptual frames" (1988b, 40). The answers to these questions have led the authors to devise a culture-specific, military-appropriate, and communication skill-centered formula for the online military culture SJTs. Four samples from the CCC course will now illustrate the design alterations that were made as a result.

SAMPLE SJTS PRODUCED FROM
THE INTENTIONAL DESIGN FORMULA

The following SJTs answer one of the four questions posed by the CCC professors as they read through the course wiki. The first represents an object familiar to all Airmen, the standard issue pocket knife. Although the skill of "perception-checking" was already included in lesson 10, the CCC student's experience in Singapore brought the skill to life in a military scenario.

I. MILITARY-APPROPRIATE OBJECT/SYMBOL = Singapore / knife

Lesson 10: Employing Effective Interaction Skills
Cuts like a Knife
   While working with foreign military equivalents in Singapore for several months at Paya Lebar Air Base, Tech Sergeant Hurston's team noticed that a few of the local Airmen were really interested in the one-handed opening knives that were standard issue for American Airmen. Since the knife is relatively inexpensive to replace and the Singaporean Airmen expressed such genuine interest in them, TSgt Hurston and several other members of the team decided to offer their knives as gifts to their Singaporean counterparts. Just before they left the country to head home, the American Airmen gave the gifts in a polite ceremonious way with both

hands as they had seen Singaporeans do over the past four months. None of the Singaporean military members would accept the gifts, even after the American Airmen adamantly insisted they take the knives.

> What should TSgt Hurston say or do next to demonstrate the cross-cultural communication skill of perception-checking?
>   a)   Present the gift again with a bow, as is expected in some South East Asian cultures to convey respect. Wait to see if the bows are returned and, if they are deeper, that is a sign that the gift is appreciated.
>   b)   Explain that he gave the gift as a token of friendship. Tell the Singaporean military members that you sensed that the gift may have offended them, and ask how they interpreted the gift.
>   c)   Explain the importance of gift-giving in American culture and that it demonstrates the amount of value that is placed on friendship.[5]

The next SJT illustrates haptics in a way that resonate with many Airmen who have been stationed in Europe. In the first few iterations of the CCC course, the majority of the relationships referred to in the lessons represented romantic or superior/subordinate relationships. However, after many students discussed the difficulty of effective communication with landlords overseas in the course wiki, the professors realized that this relationship was one that also needed attention in an intercultural context.

II. RELATIONSHIP = Italian landlord

Lesson 7: Nonverbal Communication Skills
L'APPARTAMENTO ITALIANO
   After being assigned to Aviano AB in northern Italy, Senior Airman Daniels and his wife arranged to rent an apartment owned by a friendly older Italian couple near the town of Venezia. When they dropped off their rent and deposit, the landlord's wife, who spoke English fairly well, mentioned that the American couple reminded her of them at a young age. A few days after they moved in, the landlord, who did not speak much English, brought over a bottle of wine for the new tenants. SrA Daniels took the wine bottle from the landlord and attempted to thank him in what little Italian he had picked up. The landlord, who was standing quite close to the couple, responded excitedly in Italian while waving his hands in the air. He then reached over and slapped SrA Daniels on the cheek quickly.

> Which statement offers the best nonverbal communication explanation for this behavior?
>   a)   The landlord was offended by SrA Daniels limited Italian speaking skills and was frustrated by their inability to communicate effectively. Most

Italians expect Airmen to have a functional knowledge of Italian before conducting business.

b)   The landlord was displaying affection. In Italy, slapping someone on the cheek is often seen as an act of friendship.

c)   The landlord was taken aback by the rudeness of the Americans who did not have a gift to give in return. This lack of preparedness for the initial interaction by the American is considered a sign of poor upbringing by many Italians.[6]

The third SJT acknowledges a pattern of miscommunication found in a variety of wiki contributions by Airmen who has been stationed in S. Korea. The use of silence as a form of communication was not given much attention in the CCC pilot, and after reading about students' experiences with Koreans in the course wiki, the professors felt it would be of added value to address this common American/Korean misunderstanding about the use of silence.

III. COMMUNICATION-CENTERED ACT/PRACTICE = S. Korea / Use of silence

Lesson 10: Interaction Skills
Meet in the Middle

Lee Ga-yun, a female Staff Sergeant in the ROK Air Force scheduled a meeting with SSgt Grange, a new arrival from the US Air Force, to work on a plan to complete a review of the shop's training records. When SSgt Grange arrived, SSgt Lee greeted him and they sat down.

SSgt Lee: "Good morning. I just wanted to meet with you to come up with a plan to ensure that everyone completes their training on time and that our records reflect that." She paused for a few seconds before continuing, "as we begin, do you have any questions?"

SSgt Grange: "Actually, my last shop operated very similarly so I think I should be good." He paused for a few seconds but SSgt Lee did not respond, so he then went on.

SSgt Grange: "I was the training manager there, too." A few more seconds of silence passed. "We had a great system to make sure everyone was up-to-date." A few more seconds of uncomfortable silence passed. "My supervisor there can back me up on this." SSgt Grange let another few seconds pass without a response from SSgt Lee so he decided to launch into a full explanation of his understanding of training and record-keeping procedures for almost 10 minutes, suggesting that they try the way his old shop did it first and see how that goes. He finished by asking "Does that plan sound okay to you?"

SSgt Lee paused for several moments, then responded, "It seems like your suggestion will work and we can try it out. Please let me know if I can help you in any way."

SSgt Grange left immediately feeling very frustrated with the meeting and with SSgt Lee's lack of assistance.

Based your knowledge of interaction skills, what could SSgt Grange have done to demonstrate better impression management?

a) Engage in self-monitoring. Acknowledge that he has talked a lot and allow more time for SSgt Lee to respond to his statements. Consider that he could be misreading her interaction cues and misinterpreting her silence in this conversation since it would be unlikely SSgt Lee would schedule a meeting without anything to say.

b) Manage his paralinguistic skills by using falling intonation to communicate his displeasure with SSgt Lee's silence and more forcefully end his sentences to encourage her to answer. Consider whether or not using a loud or soft volume is a more appropriate way to encourage participation in a conversation.

c) Ask more indirect questions about SSgt Lee's background to help her save face and accommodate Korea's high-context communication preferences. Consider that she might not want to contribute anything until they know each other better.[7]

Finally, after reading post-course evaluations from CCC students that requested greater variety in cultural examples throughout the course (up to that point, the majority of cultures referred to were Middle Eastern) the professors used the wiki to search out other parts of the world where students were living and working. This SJT also acknowledges that there are very few Airmen (out of the Total Force) who are pilots, and the professors wanted students to know they recognize the wide variety of Air Force professions.

IV. CULTURE-SPECIFIC LOCATION = *Bahamas* (Functional Area Manager)

Lesson 9: Identifying Communication Styles
The Communication of Respect

SMSgt Cameron Thayer is currently serving as an education and training Functional Area Manager (FAM) with the Reserves and was tasked to help run an exercise site survey and planning symposium in Freeport, Bahamas. The exercise is a joint endeavor utilizing U.S. DoD and Bahamian government agencies at multiple locations within the city of Freeport.

At each site, the facilitator, Captain Francis with the U.S. Coast Guard, calls for everyone's attention to begin introductions. He then goes around the room casually introducing participants starting first with DoD personnel and continuing with the participants from the Bahamian government. Business is addressed immediately after introductions.

SMSgt Thayer begins to notice that the head of the Bahamas' National Emergency Management Agency (NEMA) is becoming more and more agitated, impatient, and ultimately uncooperative throughout the day and brings this to the attention of the facilitator.

> Which of the following recommendations would best accommodate Bahamian communication preferences in initial interactions in order to improve the outcome of the exercise?
>
> a)   Explain that the facilitator should make formal introductions in rank order using only titles and surnames. He should introduce the Bahamian members of the team first in order of importance to communicate respect.
>
> b)   Advise the facilitator to get directly down to business; the Bahamians are becoming impatient because they feel like Captain Francis is not respecting their valuable time.
>
> c)   Suggest the possibility that Captain Francis' loud voice is being perceived as disrespectful. Recommend that the facilitator speak more softly when addressing everyone during the introductions to fit the relaxed Caribbean culture of the Bahamian archipelago.[8]

Using student wiki contributions to inform the locations, communication practices, relationships, and military-specific items of the SJTs provides an answer to the general question raised by CuDA: "How is communication shaped as a cultural practice?" (Carbaugh 2007, 168). That is, the "hubs of cultural meaning" found in the student wiki contributions (and transformed into SJTs) enhance our understanding of military students' cross-cultural experiences and can be used as both teaching and learning tools in the online classroom.

CONCLUSION

As noted by Aakhus (2007), design enterprise reveals assumptions about how communication can and should work. Including SJTs in the design process in order to privilege military students' lived experiences and subsequently updating them as the course progresses—is an example of communication best practices in distance education (Mackenzie, Fogarty, and Khachadoorian 2013). Accordingly, the work described here perpetuates the practicality of communication as an applied discipline by informing both theory and application of online intercultural communication course design. The "users" and inspiration for this project are military students whose educational journey and professional development will be enriched by the outcome of this effort.

It has been argued in this chapter that if the "practitioners" of cross-cultural competence (in this case, the military students) are to benefit from their professors' scholarly efforts, it is of the utmost importance that they be included in the instructional design process. Airmen will be unable to adopt the findings of the teaching practices intended to educate them if they do not have a say on what is being studied or understand how it impacts mission success. In the same sense, educators and researchers cannot hope to solve practical problems with their work if they don't have a clear idea of what those problems are. This is where the importance of engaged scholarship comes into play. The aim of this chapter has been to explain how student wiki contributions have informed the design of the CCC course and, in turn, how the online course design lends itself to more active student engagement. This was done by discussing the constraints and enablers of online learning, using the CuDA methodology as engaged scholars to frame the questions guiding the transformation from student wiki contribution to SJT, and suggesting a formula for intentional course design that is culture-specific, communication skill-centered, and profession-appropriate. Four military SJT examples were provided, but this formula could certainly be applied in educating any professional cohort.

The authors maintain that iterative modification adds value to intentional course design. As the quality of online courses becomes increasingly scrutinized, such efforts take on added importance by emphasizing the ways in which teaching and research as well as theory and practice are inextricably tied—for it is the nature of this connection that makes research *relevant* and teaching *transformative*. As stated by Flyvbjerg (2001, 166):

> If we want to re-enchant and empower social science . . . we must take up problems that matter to the local, national, and global communities in which we live, and we must do it in ways that matter . . . [and] we must effectively communicate the results of our research to fellow citizens. If we do this, we may successfully transform social science from what is fast becoming a sterile academic activity, which is undertaken mostly for its own sake and in increasing isolation from a society on which it has little effect and from which it gets little appreciation. We may transform social science to an activity done in times to generate new perspectives, and always to serve as eyes and ears in the future. We may, in short, arrive at a social science that matters.

The current chapter is one step taken in the direction of making social science research matter to the students it aims to serve.

NOTES

1. The interviews used as supplemental research in this chapter are based on contributions to the wiki used in the Introduction to Cross-Cultural Communication (CCC) course at the Community College of the Air Force, but no real names or direct quotations are used in order to protect the privacy of each participant.

2. SJT scores (sum of the correct responses to eight items) were positively correlated with the overall course grade which included lesson quizzes, a midterm exam, and a final exam ($r = .18$, $p < .05$, $n = 147$). The results provide construct

validity for the SJTs in that they were not only positively correlated with course grades, but also with wiki contributions ($r = .18$, $p < .01$), the knowledge post-test score ($r = .26$, $p < .01$), and the pre-flexibility score ($r = -.15$, $p < .05$). The effects of wiki participation on the Situational Judgment Test also were analyzed in a one-way analysis of variance (ANOVA). The results demonstrated significant positive main effects for participation in the wiki on the SJT scores, $F (1, 227) = 7.65$, $p < .01$. Thus, students who contributed to the wiki had significantly higher SJT scores ($M = 5.39$, $SD = 3.29$) than students who did not contribute to the wiki ($M = 4.04$, $SD = 3.21$). These results emphasize the importance of continuing to use the wiki and SJTs to improve student learning outcomes.

3. It should be noted here that students contribute to the course wiki one lesson at a time. At the end of each iteration of the course, wikis are reviewed for potential conversion into a SJT. Thus, wiki contributions in the current iteration of the course would not be (potentially) transformed into SJTs until the next iteration of the course.

4. See http://culture.af.mil/enrollmentwindow.aspx for more information.

5. "b" is the best choice. Giving a knife as a gift in Singaporean culture implies you are "severing" the friendship. In Singapore and other indirect, high-context cultures they often rely on subtle hints to convey messages that may threaten another person's "face" or cause conflict. The primary functions of communication in these cultures are to act as a social lubricant, preserve harmony, and protect face rather than simply convey information. In direct, low-context cultures gifts are just gifts and relational messages are conveyed verbally, even if it may hurt some feelings. High-context cultures often have strong gifting customs and place meaning behind every gift. TSgt Hurston displayed perception-checking by first stating why he gave the gift, then describing what he perceived was their perception, and then checking if his perception was correct.

6. "b" is the best choice. The Italian culture is a very high-contact culture in which many individuals use haptic behavior like pinching cheeks, punching shoulders, and slapping as a friendly gesture. Many older Italians do this with younger people, especially family members like children, grandchildren, nieces, and nephews. These exuberant nonverbal behaviors may be rough or invasive by American standards but they are quite common among Italians. The other choices reflected cultural approaches to linguistic competence and gifting rather than nonverbal communication.

7. "a" is the best choice. Successfully interpreting turn-taking cues in Korea often requires accepting a moment of silence lasting between five and ten seconds between speaking turns, which can feel very awkward to Americans who may continue to speak before their Korean counterpart perceives it is their turn. Koreans wait for this length of silence to indicate that it is their turn to speak and may not be comfortable interrupting a speaking partner to share their ideas. SSgt Grange is practicing Impression management here by acknowledging that he has talked a lot (self-monitoring) and correctly associating SSgt Lee's silence with cultural preferences rather than inability to contribute (perceptual acuity).

8. "a" is the best choice. The best accommodation of Bahamian introduction preferences is to make formal introductions using only titles and surnames and address the Bahamian members of the team first in order of importance to communicate respect. The Bahamian culture is very hierarchical and first names are typically used only by very close friends and family. Once the facilitator began to

introduce the Bahamians first, the NEMA Minister's body language and attitude changed and he became more engaged in the plan.

REFERENCES

Aakhus, Mark. "Communication as Design." *Communication Monographs* 74, no. 1 (2007): 112–117. doi:10.1080/03637750701196383.

Allen, I. Elaine, and Jeff Seaman. *Changing Course: Ten Years of Tracking Online Education in the United States.* Oakland, CA: Babson Survey Research Group and Quahog Research Group, 2013. February 6, 2015. http://www.onlinelearning-survey.com/reports/changingcourse.pdf.

Alred, Gerald J. "Bridging Cultures: The Academy and the Workplace." *Journal of Business Communication* 43, no. 2 (2006): 79–88. doi:10.1177/0021943605285659.

American Society for Training and Development. "Less Classroom, More Technology." *Training and Development,* May 2005.

Ang, Soon, and Linn Van Dyne. *Handbook of Cultural Intelligence: Theory, Measurement, and Applications.* Armonk, NY: M.E. Sharpe, 2008.

Artino, Anthony R. "Motivational Beliefs and Perceptions of Instructional Quality: Predicting Satisfaction with Online Training."*Journal of Computer Assisted Learning* 24 (2008): 260–70. doi:10.1111/j.1365-2729.2007.00258.x.

Barrera, Isaura, and R. M. Corso. "Cultural Competency as Skilled Dialogue." *Topics in Early Childhood Special Education* 22, no. 2 (2002): 103–13. doi:10.1177/02711214020220020501.

Beckett, Gulbahar H., Carla Amaro-Jiménez, and Kelvin S. Beckett. "Students' Use of Asynchronous Discussions for Academic Discourse Socialization." *Distance Education* 31, no. 3 (2010): 315–35. doi:10.1080/01587919.2010.513956.

Beldarrain, Yoany. "Distance Education Trends: Integrating New Technologies to Foster Student Interaction and Collaboration." *Distance Education* 27, no. 2 (2006): 139–153.

Ben-Yoav Nobel, Orly, Brian Wortinger, and Sean Hannah. "Winning the War and the Relationships: Preparing Military Officers for Negotiations with Non-Combatants." 2007. www.dtic.mil/cgi-bin/GetTRDoc?AD=ADA472089. U.S. Army Research Institute for the Behavioral and Social Sciences. Research Report 1877. Arlington, Virginia.

Blackboard, Inc. "Blackboard Exemplary Course Program Rubric." Blackboard.com. 2012. Accessed December 11, 2014. http://www.blackboard.com/getdoc/7deaf501-4674-41b9-b2f2-554441ba099b/2012-Blackboard-Exemplary-Course-Rubric.aspx.

Blackboard, Inc. "Wikis." Blackboard Help. Accessed August 12, 2013. https://help.blackboard.com/en-us/Learn/9.1_SP_10_and_SP_11/Instructor/050_Course_Tools/Wikis#.

Blankenship, Mark. "How Social Media Can and Should Impact Higher Education." *Education Digest* 76, no. 7 (March 2011): 39–42.

Botturi, Luca. "Design Models as Emergent Features: An Empirical Study in Communication and Shared Mental Models in Instructional Design." *Canadian Journal of Learning and Technology* 32, no. 2 (Spring 2006). Accessed February 6, 2015. http://www.cjlt.ca/index.php/cjlt/article/view/50/47.

Bradford, Peter, Margaret Porciello, Nancy Balkon, and Debra Backus. "The Blackboard Learning System: The Be All and End All in Educational Instruction?" *Journal of Educational Technology Systems* 35, no. 3 (2007): 301–14. doi:10.2190/X137-X73L-5261-5656.

Branzburg, Jeffrey, and Kristen Kennedy. "Online Professional Development." *Technology and Learning* 22, no. 2 (2001): 18–27.

Brunk-Chavez, Beth L., and Shawn J. Miller. "Decentered, Disconnected, and Digitized: The Importance of Shared Space." *Kairos* 11, no. 1 (Spring 2007). Accessed December 21, 2014. http://kairos.technorhetoric.net/11.2/topoi/brunk-miller/decenteredtest/decentered.pdf.

Bryant, Susan. "The Five Habits: Building Cross-cultural Competence in Lawyers." *Clinical Law Review*, Fall 2001. Accessed February 6, 2015. http://www.law.cuny.edu/faculty/pedagogy/bryant-habits.pdf.

Bryce, Nadine. "Teacher Candidates' Collaboration and Identity in Online Discussions." *Journal of University Teaching and Learning Practice* 11, no. 1 (2014). Accessed February 6, 2015. http://ro.uow.edu.au/jutlp/vol11/iss1/7/.

Campion, Michael C., Robert E. Ployhart, and William I. MacKenzie, Jr. "The State of Research on Situational Judgment Tests: A Content Analysis and Directions for Future Research." *Human Performance* 27, no. 4 (2014). doi:10.1080/0895928 5.2014.929693.

Carbaugh, Donal. "Comments on 'Culture'; in Communication Inquiry." *Communication Reports* 1, no. 1 (1988b): 38–41. doi:10.1080/08934218809367460.

Carbaugh, Donal. "Toward a Perspective on Cultural Communication and Intercultural Contact." *Semiotica* 80, no. 1–2 (1990): 15–36. doi:10.1515/semi.1990.80.1-2.15.

Carbaugh, Donal. "Cultural Discourse Analysis: Communication Practices and Intercultural Encounters." *Journal of Intercultural Communication Research* 36, no. 3 (2007): 167–182. doi:10.1080/17475750701737090.

Carbaugh, Donal. "Ethnography of Communication." In *The International Encyclopedia of Communication*, edited by Wolfgang Donsbach. Hoboken, NJ: Wiley-Blackwell, 2008.

Carbaugh, Donal, and Sally O. Hastings. "A Role for Communication Theory in Ethnography and Cultural Analysis." *Communication Theory* 2, no. 2 (1992): 156–65. doi:10.1111/j.1468-2885.1992.tb00035.x.

Carbaugh, Donal, Elena V. Nuciforo, Makoto Saito, and Dong-Shin Shin. "'Dialogue' in Cross-Cultural Perspective: Japanese, Korean, and Russian Discourses." *Journal of International and Intercultural Communication* 4, no. 2 (2011): 87–108. doi:10.1080/17513057.2011.557500.

Center for Instructional & Learning Technologies. Blackboard Tips for Faculty. Accessed June 16, 2010. http://und.edu/academics/cilt/instructional-design/weekly-tips/weeklytipsjune16.pdf.

Chandler, Jennifer V. "Why Culture Matters: An Empirically-based Pre-deployment Training Program." Master's thesis, Naval Postgraduate School, 2005. Accessed February 6, 2015. http://www.dtic.mil/dtic/tr/fulltext/u2/a439382.pdf.

Christian, Michael S., Bryan D. Edwards, and Jill C. Bradley. "Situational Judgment Tests: Constructs Assessed and a Meta-Analysis Of Their Criterion-Related Validities." *Personnel Psychology* 63, no. 1 (2010): 83–117. doi:10.1111/j.1744-6570.2009.01163.x.

Cleaver, Samantha. "Beyond Blackboard and into Virtual Communities." *Diverse: Issues in Higher Education* 25, no. 18 (October 2008): 32. Accessed December 27, 2014. ProQuest.

Cox, Taylor. *Cultural Diversity in Organizations: Theory, Research and Practice.* San Francisco: Berrett-Koehler, 1994.

Crandall, Sonia J., Gavin George, Gail S. Marion, and Stephen Davis. "Applying Theory to the Design of Cultural Competency Training for Medical Students: A Case Study." *Academic Medicine* 78, no. 6 (2003): 588–94.

Crosson, Jesse C., Weiling Deng, Chantal Brazeau, Linda Boyd, and Maria Soto-Greene. "Evaluating the Effect of Cultural Competency Training on Medical Student Attitudes." *Family Medicine* 36, no. 3 (March 2004): 199–203. Accessed February 6, 2015. http://www.stfm.org/fmhub/fm2004/March/Jesse199.pdf.

Dringus, Laurie P., and Timothy Ellis. "Temporal Transitions in Participation Flow in an Asynchronous Discussion Forum." *Computers & Education* 54, no. 2 (2009): 340–49. doi:10.1016/j.compedu.2009.08.011.

Earley, Christopher P., and Soon Ang. *Cultural Intelligence: Individual Interactions across Cultures.* Stanford, CA: Stanford University Press, 2003.

Fenton, Celeste, and Brenda W. Watkins. "Online Professional Development for K–12 Educators: Benefits for School Districts with Applications for Community College Faculty Professional Development." *Community College Journal of Research and Practice* 31, no. 6 (2007): 531–33. doi:10.1080/10668920701359946.

Flyvbjerg, Bent. *Making Social Science Matter: Why Social Inquiry Fails and How It Can Succeed Again.* Oxford, UK: Cambridge University Press, 2001.

Grbich, Carol. *Qualitative Data Analysis: An Introduction.* London: SAGE Publications, 2007.

Greene-Sands, Robert, and Allison Greene-Sands, eds. *Cross-Cultural Competence for a 21st Century Military.* Lanham, MD: Lexington, 2014.

Gustafson, Kent L., and Robert M. Branch. *Survey of Instructional Development Models,* 4th edition. 2002.

Hara, Noriko, Curtis J. Bonk, and Charoula Angeli. "Content Analysis of Online Discussion in an Applied Educational Psychology Course." *Instructional Science* 28, no. 2 (2000): 115–52. doi:10.1023/A:1003764722829.

Hew, Khe F., and Wing Sum Cheung. "Students' Use of Asynchronous Voice Discussion in a Blended Learning Environment: A Study of Two Undergraduate Classes." *Electronic Journal of E-Learning* 10, no. 4 (2012): 360–67. www.ejel.org/issue/download.html?idArticle=215.

Hew, Khe Foon, Wing Sum Cheung, and Connie Siew Ling Ng. "Student Contribution in Asynchronous Online Discussion: A Review of the Research and Empirical Exploration." *Instructional Science* 38, no. 6 (2010): 571–606. doi:10.1007/s11251-008-9087-0.

Holsombach-Ebner, Cinda. "Quality Assurance in Large Scale Online Course Production." *Online Journal of Distance Learning Administration* 16, no. 2 (September 1, 2013). Accessed February 6, 2015. http://www.westga.edu/~distance/ojdla/fall163/holsombach-ebner164.html.

Holzer, Elie. "Professional Development of Teacher Educators in Asynchronous Electronic Environment: Challenges, Opportunities and Preliminary Insights from Practice." *Educational Media International* 41, no. 1 (2004): 81–89. doi:10.1080/0952398032000105139.

Jeffreys, Marianne. "Dynamics of Diversity: Becoming Better Nurses through Diversity Awareness." *NSNA Imprint*, November/December 2008, 36–41. Accessed February 6, 2015. http://www.nsna.org/Portals/0/Skins/NSNA/pdf/Imprint_NovDec08_Feat_Jeffreys.pdf.

Katzenstein, Mary F., and Judith Reppy. *Beyond Zero Tolerance: Discrimination in Military Culture*. Lanham, MD: Rowman and Littlefield Publishers, 1999.

Kelly, Rob. "Building Community in Self-paced Online Courses." *Online Cl@ssroom*, 1 (June 2005): 7. Accessed February 6, 2015. http://www.magnapubs.com/newsletter/online-classroom/16/building_community_in_self_paced_online_courses-10211-1.html.

Kenny, Richard F., Zuochen Zhang, Richard A. Schwier, and Katy Campbell. "A Review of What Instructional Designers Do: Questions Answered and Questions Not Asked." *Canadian Journal of Learning and Technology* 31, no. 1 (Winter 2005): 9–16. Accessed February 6, 2015. http://www.cjlt.ca/index.php/cjlt/article/view/147/140.

Krumm, Stefan, Filip Lievens, Joachim Hüffmeier, Anastasiya A. Lipnevich, Hanna Bendels, and Guido Hertel. "How 'Situational' Is Judgment in Situational Judgment Tests?" *Journal of Applied Psychology* (August 11, 2014): 1–18. doi:10.1037/a0037674.

Kuhlenschmidt, Sally. "Why Use Blogs and Wikis in Blackboard?" Lecture, January 21, 2009. Accessed February 6, 2015. https://www.wku.edu/teaching/online/blackboardblogswiki.ppt.

Lackey, Dundee. "Why Wiki?" *Kairos Praxis Wiki* 12, no. 1 (2007). Accessed February 6, 2015. http://praxis.technorhetoric.net/tiki-index.php?page=Why_Wiki.

Larson, David K., and Chung-Hsien Sun. "Comparing Student Performance: Online Versus Blended Versus Face-to-Face." *Journal of Asynchronous Learning Networks* 13, no. 1 (2009): 31–42. Accessed February 6, 2015. https://idt7895.files.wordpress.com/2009/05/comparing-student-performance-in-different-delivery-methods.pdf.

Lave, Jean, and Etienne Wenger. *Situated Learning: Legitimate Peripheral Participation*. Cambridge: Cambridge University Press, 1991.

Lee, Silvia W., and Chin-Chung Tsai. "Identifying Patterns of Collaborative Knowledge Exploration in Online Asynchronous Discussions." *Instructional Science* 39, no. 3 (May 2011): 321–47. doi:10.1007/s11251-010-9131-8.

Leung, Kwok, Soon Ang, and Mei Ling Tan. "Intercultural Competence." *Annual Review of Organizational Psychology and Organizational Behavior* 1, no. 1 (2014): 489–519. doi:10.1146/annurev-orgpsych-031413-091229.

Levin, Barbara B., Ye He, and Holly H. Robbins. "Comparative Analysis of Preservice Teachers' Reflective Thinking in Synchronous Versus Asynchronous Online Case Discussions." *Journal of Technology and Teacher Education* 14, no. 3 (2006): 439–60. libres.uncg.edu/ir/uncg/f/B_Levin_Comparative_2006.pdf.

Licona, Miguel M., and Binod Gurung. "Asynchronous Discussions in Online Multicultural Education." *Multicultural Education* (Fall 2011): 2–8. Accessed February 6, 2015. http://files.eric.ed.gov/fulltext/EJ986884.pdf.

Lin, Qiuyun. "Student Satisfactions in Four Mixed Courses in Elementary Teacher Education Program." *Internet and Higher Education* 11, no. 1 (2008): 53–59. Accessed February 6, 2015. http://eric.ed.gov/?id=EJ796651.

Lord, Gillian, and Lara L. Lomicka. "Developing Collaborative Cyber Communities to Prepare Tomorrow's Teachers." *Foreign Language Annals* 37, no. 3 (2004): 401–08. doi:10.1111/j.1944-9720.2004.tb02698.x.

Lorenzetti, Jennifer P. "Lessons Learned about Student Issues in Online Learning." *Distance Education Report* 9, no. 18 (2005): 1–7.

Mackenzie, Lauren. "Strategic Enablers: How Intercultural Communication Skills Advance Micro-Level International Security." *Journal of Culture, Language & International Security* 1, no. 1 (Summer 2014). Accessed February 6, 2015. http://culture.af.mil/library/pdf/MackenzieJCLISMay2014.pdf.

Mackenzie, Lauren, Patricia Fogerty, and Angelle Khachadoorian. "A Model for Military Online Culture Education: Key Findings and Best Practices." *EDUCAUSE Review* 48, no. 4 (July/August 2013). Accessed February 6, 2015. http://www.educause.edu/ero/article/model-online-military-culture-education-key-findings-and-best-practices.

Mackenzie, Lauren, and Megan Wallace. "Distance Learning Designed for the U.S. Air Force." *Academic Exchange Quarterly* 16, no. 2 (Summer 2012): 55–60.

Mackenzie, Lauren, and Megan Wallace. "Cross-Cultural Communication Contributions to Professional Military Education: A Distance Learning Case Study." In *Cross-Cultural Competence for a 21st Century Military: Culture, the Flipside of COIN*, edited by Robert R. Greene-Sands and Allison Greene-Sands, 239–58. Lanham, MD: Lexington Books, 2014.

Manning, Michelle. "Building Presence and Community in the Online Class." *Kairos: Rhetoric, Technology, and Pedagogy*, February 27, 2007. Accessed February 6, 2015. http://praxis.technorhetoric.net/tiki-index.php?page=Building_Presence.

May, Larissa, Kimberly D. Acquaviva, Annette Dorfman, and Laurie Posey. "Medical Student Perceptions of Self-Paced, Web-Based Electives: A Descriptive Study." *American Journal of Distance Education* 23, no. 4 (2009): 212–23. doi:10.1080/08923640903332120.

McDaniel, Michael A., Deborah L. Whetzel, and Nhung T. Nguyen. *Situational Judgment Tests for Personnel Selection*. In *An IPMAAC Personel Assessment Monograph*, edited by Leilani Yan. Alexandria, VA: IPMA-HR Assessment Council, 2006.

Milburn, Trudy. *Nonprofit Organizations: Creating Membership through Communication*. Cresskill, NJ: Hampton Press, 2009.

Miller, Derick, and Lisa Rudnick. *The Security Needs Assessment Protocol: Improving Operational Effectiveness through Community Security*. Report. New York and Geneva: United Nations Publications, 2008.

Miller, Katherine. *Organizational Communication: Approaches and Processes*, 6th edition. Boston, MA: Wadsworth, 2006.

Molinsky, Andrew L. "The Psychological Processes of Cultural Retooling." *Academy of Management Journal* 56, no. 3 (2013): 683–710. doi:10.5465/amj.2010.0492.

Moneta, Giovanni B., and Synnöve S. Kekkonen-Moneta. "Affective Learning in Online Multimedia and Lecture Versions of an Introductory Computing Course." *Educational Psychology* 27, no. 1 (2007): 51–74. doi:10.1080/01443410601061413.

Murphy, Elizabeth. "Recognising and Promoting Collaboration in an Online Asynchronous Discussion." *British Journal of Educational Technology* 35, no. 4 (2004): 421–31. doi:10.1111/j.0007-1013.2004.00401.x.

Murphy, Elizabeth, María A. Rodríguez-Manzanares, and Michael Barbour. "Asynchronous and Synchronous Online Teaching: Perspectives of Canadian

High School Distance Education Teachers." *British Journal of Educational Technology* 42, no. 4 (2011): 583–91. doi:10.1111/j.1467-8535.2010.01112.x.

NCDAE. "Tips and Tools: Content Management Systems and Accessibility." *Content Management Systems & Accessibility*. June 2006. Accessed January 22, 2015. http://ncdae.org/resources/factsheets/cms.php.

Nelson, Edwin B. "Cultural Awareness: Resources Can Help Prepare Soldiers before Deployments." *Infantry Magazine*, January/February 2007.

Pacanowsky, Michael E., and Nick O'Donnell-Trujillo. "Organizational Communication as Cultural Performance." *Communication Monographs* 50, no. 2 (1983): 126–47. doi:10.1080/03637758309390158.

Philipsen, Gerry. *Speaking Culturally: Explorations in Social Communication*. Albany, NY: State University of New York Press, 1992.

Picciano, Anthony G. "Beyond Student Perceptions: Issues of Interaction, Presence, and Performance in an Online Course." *Journal of Asynchronous Learning Networks* 6, no. 1 (2002): 21–40. doi:10.1.1.98.6506.

Poyner, Robert D. "The Hyphenated Airman." *Air and Space Power Journal* 21, no. 3 (Fall 2007): 23–25. Accessed February 6, 2015. http://www.au.af.mil/au/afri/aspj/airchronicles/apj/apj07/fal07/fal07.pdf.

Rempel, Hannah G., and Paula S. McMillan. "Using Courseware Discussion Boards to Engage Graduate Students in Online Library Workshops." *Internet Reference Services Quarterly* 13, no. 4 (2008): 363–80. doi:10.1080/10875300802326350.

Restauri, Sherri L. "Creating an Effective Online Distance Education Program Using Targeted Support Factors." *TechTrends* 48, no. 6 (2004): 32–39. doi:10.1007/BF02763580.

Rivait, Jessica. "Course Management Systems as Ongoing Classroom Memory: ANGEL as a Living Archive." *Kairos: A Journal of Rhetoric* 15, no. 1 (July 11, 2014). Accessed February 6, 2015. http://praxis.technorhetoric.net/tiki-index.php?page=Classroom_Memory.

Roberts, Barbara. "Interaction, Reflection and Learning at a Distance." *Open Learning: The Journal of Open and Distance Learning* 17, no. 1 (2002): 39–55. doi:10.1080/02680510120110166.

Rockstuhl, Thomas, Soon Ang, Kok-Yee Ng, Filip Lievens, and Linn VanDyne. "Putting Judging Situations into Situational Judgment Tests: Evidence from Intercultural Multimedia SJTs." *Journal of Applied Psychology* (October 6, 2014). Accessed February 6, 2015. http://dx.doi.org/10.1037/a0038098.

Rockstuhl, Thomas, A. Presbitero, Kok-Yee Ng, and Soon Ang. "Metacognitive Cultural Intelligence and Offshoring Performance: Predictive Validity of an Intercultural Situational Judgment Test (iSJT)." Presentation at *The 3rd International Conference on Outsourcing of Information Services*. Proceedings, Mannheim, Germany. June 10–11, 2013.

Rowley, Jennifer, and Jennifer O'Dea. "How Do Students Perceive the Enhancement of Their Own Learning? A Comparison of Two Education Faculties' Experiences in Building an Online Learning Community for Bachelor of Music Education and Bachelor of Education Students." In *Australian Teacher Education Association Annual Conference Proceedings Archive*. Proceedings of "Teacher Education Crossing Borders: Cultures, Contexts, Communities and Curriculum" the Annual Conference of the Australian Teacher Education Association (ATEA), Albury (July 2009). Accessed February 6, 2015. http://atea.edu.au/ConfPapers/2009/Refereed/Rowley.pdf.

Sandars, John, and Michelle Langlois. "Online Learning Networks for General Practitioners: Evaluation of a Pilot Project." *Education for Primary Care* 16 (2005): 688–96.

Santovec, Mary Lou. "Doing Online Professional Development—Online." *Distance Education Report* (September 15, 2004): 6–7.

Short, Nancy M. "Asynchronous Distance Education." *Technological Horizons in Education Journal* 28, no. 2 (2000). Accessed February 6, 2015. http://thejournal.com/magazine/vault/A3001.cfm.

Snow, Eleanour, and Perry Sampson. Course Platforms for Teaching Online. Authored as Part of the 2010 Workshop, Teaching Geoscience Online—A Workshop for Digital Faculty, 2010. Accessed February 6, 2015. http://serc.carleton.edu/NAGTWorkshops/online/platforms.html.

Spencer, Lyle M., and Signe M. Spencer. *Competence at Work: Models for Superior Performance*. New York: Wiley, 1993.

Swan, Karen. "Learning Effectiveness: What the Research Tells Us." In *Elements of Quality Online Education: Practice and Direction*, edited by John R. Bourne and Janet C. Moore, 13–47. Needham, MA: Sloan Consortium, 2003.

Szabady, Gina, Crystal N. Fodrey, and Celeste Del Russo. "Digital (Re)Visions: Blending Pedagogical Strategies with Dynamic Classroom Tactics." *Kairos* 19.1 (Fall 2014).

Teasley, Martell L. "Perceived Levels Of Cultural Competence Through Social Work Education And Professional Development For Urban School Social Workers."*Journal of Social Work Education* 41, no. 1 (2005): 85–98. doi:10.5175/JSWE.2005.200300351.

Van De Ven, Andrew H. *Engaged Scholarship: A Guide for Organizational & Social Research*. Oxford: Oxford University Press, 2007.

Varvel, Todd K. "Ensuring Diversity Is Not Just another Buzz Word." Master's thesis, Air University, 2000. Accessed February 6, 2015. http://www.au.af.mil/au/awc/awcgate/acsc/00-180.pdf.

Weingardt, Kenneth R., Michael A. Cucciare, Christine Bellotti, and Wen Pin Lai. "A Randomized Trial Comparing Two Models of Web-based Training in Cognitive–Behavioral Therapy for Substance Abuse Counselors." *Journal of Substance Abuse Treatment* 37, no. 3 (2009): 219–27. doi:10.1016/j.jsat.2009.01.002.

Witteborn, Saskia. "Discursive Grouping in a Virtual Forum: Dialogue, Difference, and the 'Intercultural.'" *Journal of International and Intercultural Communication* 4, no. 2 (2011): 109–26. doi:10.1080/17513057.2011.556827.

Xia, Jianhong, John Fielder, and Lou Siragusa. "Achieving Better Peer Interaction in Online Discussion Forums: A Reflective Practitioner Case Study." *Issues in Educational Research* 23, no. 1 (2013): 97–113. February 6, 2015. http://www.iier.org.au/iier23/xia.pdf.

Zha, Shenghua, and Christy L. Ottendorfer. "Effects of Peer-Led Online Asynchronous Discussion on Undergraduate Students' Cognitive Achievement." *American Journal of Distance Education* 25, no. 4 (2011). doi: DOI:10.1080/08923647.2011.618314.

Epilogue

Implications for Improving UX Practice
Trudy Milburn and James L. Leighter

This book explored several representative local strategies studies about user experiences with digital media. Our aim has been to illustrate the ways in which this set of methods has been fruitfully employed to enhance design work. The progression from micro-interaction analysis to more macro-interpretations about relationships and culture have demonstrated several moments within the design process where cultural context is used to gain a more refined understanding about digital interaction through everyday situations.

In these concluding remarks, we focus on some themes that several chapters have touched upon, raising further questions about the use of local strategies research (LSR) and the implications for practice in design settings. We will also reflect upon the ways LSR is being articulated and make some suggestions for further refinement as we continue this applied work. Finally, we will offer some next steps that can be taken by practitioners who want to embed these suggestions into their own work.

THEMATIC REVIEW: LSR TOOLKIT

Each chapter provides a general orientation to methods that we are grouping together as "local strategies research." Using previously defined labels of ethnography of communication, cultural discourse theory, and cultural codes theory, we have compiled what can be considered a toolbox of analytic methods researchers employ to help interpret the actions and patterned practices witnessed through participant observation and supplemented by interviews. Digital media allow researchers to record or take screen shots to preserve the moment-to-moment interactions that have long been difficult to capture during face-to-face encounters. Local strategies researchers rely upon these recordings as well as their field notes to conduct their more fine-grained analyses. Following the analysis and interpretation, researchers are able to reflect on the ways that designs shape communicative situations in interaction, and the ways in which people both respond to, and co-construct, those situations.

Although the project of cultural discourse analysis has been more or less focused on five specific modes of research: theoretical, descriptive, interpretive, comparative, and critical, LSR adds a sixth: application. By moving to application after findings based on the first five modes of research, we can ask how developers and designers can alter or change the digital media through which a practice is being achieved in a way that enhances the experience for those who employ the designs. Although these five modes seem linear, by adding the sixth, we come full circle inching back to theoretical questions (see diagram from introduction) about the moments in the design process when local strategies research can provide additional resources for consideration.

It is the project of application that moves into the realm of lived experience in a way that necessitates seeing this research as a momentary snapshot of an ever changing array of devices and digital media entering our world. The discrete moment-in-time not only must be accounted for, but historic considerations must be considered as new digital media become part of any interaction. Systematic and rigorous attention to the constantly changing digital landscape leads us, perhaps, to make smaller, incremental claims, and may lead us to consider our relationship with the participants we examine a bit differently. Regarding this latter point, we bring into question pre-formulated notions of users by developing more nuanced understandings of persons-in-interaction.

LSR can be considered an "applied" extension of the aforementioned theories as its commitments are grounded in two key areas (a) participants' actual practices (in this case, the common use of digital media); and (b) the reference to cultural comparisons that can provide a way to recognize and differentiate what is common among practices and what differentiates the practices of the particular group being investigated. The contribution that LSR makes to other methods is to purposefully consider how interactions with or through digital media may be improved. In reading the studies recounted herein, we have become more sensitized to how different situations and cultural norms provide opportunities to continually learn and challenge what we know about communicative practices, and to consider the application to new situations and cultural contexts.

Dividing this volume into three sections momentarily affords a focus on one particular dimension of research: the smallest units comprised of communicative acts and patterned practices, the ways specific communicative acts are used to form and maintain relationships, and the way identities are exhibited through practices that often demonstrate larger cultural values and beliefs. Such a division provides analytic perspective while also maintaining a commitment to the ways in which these dimensions are inextricably bound together in the moment-by-moment unfolding of social interaction.

The authors have explored users' interactions within particular scenarios or communicative situations. In the first section, we focused on some of the activities that are performed with digital media within two particular contexts: in-car interaction and online training. Through these chapters, we learned more about particular normative practices and consequences of breaches or breakdowns that occur when using new digital media to communicate. The authors have illustrated different ways to conceptualize the nature and origins of common or likely misunderstandings. For example, from Molina-Markham et al. we learned that the boundaries of conversational openings and closings were not as easily understood by the in-car system as drivers anticipated. From Hart, we learned that the use of the tutorial interface that demanded a strict adherence to a conversational script impeded breaks in that script. In both cases, participants had to negotiate what they knew about typical conversational sequences and the constraints they encountered using the new digital medium.

In part II, each of the authors explored how relationships are built and maintained, or may suffer and decline, when conducting interaction through particular digital media. The contexts that formed the base of these studies ranged from formal business meetings to more informal social encounters. While the purpose or goals for interaction vary in the selection of a given device or medium, for instance, a mobile application that easily facilitates social interactions rather than a software rubric application that evaluates students in a classroom, we find commonality in that the studies illustrate the way digital media can help or hinder the joint accomplishment of shared goals. The two ways to conceptualize relationships, goal completion versus socialization were either enhanced or constrained by the new digital medium used in each case.

In the third part, Poutiainen (chapter 6) and Mackenzie and Wallace (chapter 7) described research about user identities grounded in particular places. At issue within these chapters are the possible and relevant identities that emerge when interacting with and through digital media, and often those that transcend place. In fact, while not explicitly the focus of the previous two sections, we find this same theme recurrent in the other chapters as well. By using a cultural identity lens, we learned that it is consequential for Finns to frequently talk on cell phones and to have a public, historic recognition of themselves as reticent speakers. This case would form just one scenario within Mackenzie and Wallace's online intercultural course, where the focus is on the use of situationally-embedded prompts to help provide military members context for anticipated professional encounters in multiple cultural contexts.

Overall, the chapters have made the assumption that both designers and LSR theorists recognize the importance of context when conducting research and making interpretive claims. After considering the particular contexts and specific scenarios contained herein, we are left with several questions. We can examine these questions in new situations on a quest to continually explore the relationship between people, their actions, and the influence of the digital media they use.

- For what purpose was the particular digital medium chosen?
- Was the digital medium (re)evaluated or simply accepted as something one should continue using? Did its affordances or strengths become the focus of continued interaction or does its use become taken for granted?
- What was culturally distinct about the actions or practices of users? Did the display of identity rely upon a place-bounded group, or did the enacted and performed identities create culture in others ways?
- What is the relationship between an individual user or person's activities to others, a collective task or goal, and the digital medium itself?
- How do the users actively participate in the realized *in situ* deployment of a design?

As was introduced in the introduction, a typical design process includes many moments for designer-user interaction including defining local problematics, user-interaction analysis of prototypes or fully formed designs and, we posit, opportunities to reflect on design use after it has been deployed for some period of time. One's ability to accurately interpret data as meaningful for participants themselves is one of the largest hurdles. The meaning of a particular communicative act varies by situation. That is why a comparison of similar interactions across a variety of situations is key. This is often characterized as a literature review in scholarly research papers, and it provides one with a range of potential interpretations based on previous studies. Applying this cross-cultural knowledge to build and enhance digital media can ensure that tools intended for users in different cultural contexts can be meaningfully used in others. By conducting LSR research we may learn things about users that we cannot act upon due to constraints (time, money, or other resources and priorities). The question becomes, should we take the time to learn these things? And, if we commit to doing so, how will what is learned shape the design itself? One could argue that researchers always learn more than they can productively apply to a given situation. However, the idea of contributing to the stock of knowledge to make comparisons among ways of speaking was Hymes's (1964) original intent. Furthermore, he provided guidelines to compare ethnographic findings (SPEAKING as but one example). While some may discount the use of ethnographic methods, such as participant observation, as too time consuming and expensive for business and production needs to be met, we feel that their benefits outweigh their costs.

From the perspective of the user, we are always learning and adapting to technological enhancements from one product to another, from one scenario to another. A customer service encounter via telephone produces expectations about the sequence of events and topics that are then potential fodder for the customer service experience online via synchronous chat. Taking into account rapidly changing expectations based on the use of many different digital media is one challenge we face today. We believe it is a challenge that can be addressed by helping to transfer or translate known conversational sequences, findings about relationship and culture into new situations and cultural contexts, as well as into new digital media.

The way digital media impacts relationships is another significant area to address. Several studies illustrated the ways role misunderstandings can arise, especially between traditional roles such as teachers/trainers and students/peers (see chapters 1, 4, 5, and 7). The ways digital media may transcend hierarchy or disconnect people who were formerly part of a team (chapters 5 and 3 respectively) may be an unexpected results or consequence of new digital media for which we should account. However, as Bouwmeester describes in chapter 4, merely reacting to this knowledge as a problem to fix or re-design may miss more nuanced cultural meaning related to a particular communicative breach. Rather than quickly move to address these issues with system modifications or enhancements, we would advocate learning more about the historic, cultural meanings upon which these interpretations are based.

METHODS OF INVESTIGATION

We have alluded to an LSR toolkit and now turn to the question, what do the methods used to investigate user experiences provide? When discussing methods it is often useful to separate the data collection from the data analysis. For data collection, several authors conducted participant observation, with some writing field notes and others including audio recordings of interaction. Some authors conducted interviews as their primary method and others conducted interviews as a supplemental area of investigation. For those whose primary data was initially produced through a digital medium, some have captured screen shots of interaction. When collecting data, we can interrogate how face-to-face interview data compares with recorded talk conducted entirely online. These questions are beyond the scope of this volume; however, researchers should always ask, how do our methods impact what we can learn about "local strategies" for acting, for relating, and for being?

Rather than beginning with particular personas as one might expect from a traditional design perspective, we have begun with communicative acts and patterned practices. This follows a trend in ethnography of communication research that departs from Hymes's original suggestion to begin with a speech community, where we have witnessed an increasing trend to begin with a communicative practice (see Milburn 2015). Briefly this is the case because people display multiple and overlapping identities and the distinctions that were once based on common ethnic or national identifiers no longer are the tightly drawn boundaries they were once thought to be.

Revisions to Hymes's original categories continue to be shaped by new research. In fact, consider Hymes's original category of communicative "channel." When first described, the current variety of digital media did not exist. Hymes, however, did anticipate that the use of or meanings for any given channel did vary culturally.

> There is a tendency to take the value of a channel as given across cultures, but here, as with every aspect and component of communication, the value is problematic and requires investigation. (Consider for example the specialization of writing to courtship among young people by the Hanunoo, and to a borrowed religion among the Aleut; and the complex and diverse profiles with regard to the role of writing in society, and in individual communicative events, for traditional Chinese, Korean, and Japanese cultures, with regard both to the Chinese texts shared by all and to the materials specific to each.) To provide a better ethnographic basis for the understanding of the place of alternative channels and modalities in communication is indeed one of the greatest challenges to studies of the sort we seek to encourage. At the same time, such work, whether on channels or some other aspect and component, profits from taking into account the complete context of the activity of the system of communication of the community as a whole. (Hymes 1964, 25)

Based on the combined findings from this volume, Hymes's notion of channel is beginning to be refined to include digital media and this, in turn, will impact the way these methods are employed in future studies. Some of the ways we observe traditional notions of channel changing is the (a) multiple channels available for use, as well as used, during any one situation; (b) the way channel affords and constrains communicative acts such as scripts, laughter, etc.; and (c) the way channel may be made invisible or taken for granted in ways unanticipated in the past.

Finally, these methods have shown particular promise for digital media design. Processes and cycles of product development vary, but often teams of developers and designers must make alterations to the interface at several discrete moments during prototyping. Molina-Markham et al.'s research (see chapter 1) illustrates the potential for creating adaptive technologies based on actual interactions. Their research provides a window into the responses an in-car audio system makes to a user's commands. If we consider the repeated communicative acts that were observed, we can imagine a system that is more responsive to particular requests that are made based on previous information learned. A move in the direction of creating automatically adaptive systems still requires close consideration of the ways people actually interact, what they do when breaches occur, and how these interactions form the basis of stronger or more adversarial relationships.

LEARNING AND UX

Since this volume includes three chapters explicitly about the context of online learning, it is fair to ask, what is the relationship between learning and digital media? Indeed, one might go so far as to suggest that all of the chapters included in this volume have implications for learning—in that a new digital medium must be learned in each case. Learning is often the label given to the demonstration of the "right" (for any given community) way of doing something. However, it can also be used as a label for indicating any change that occurs over time (often, as long as that change is considered "positive" by the cultural group or community who is responsible for its evaluation).

Within the realm of education, there are a number of ways digital media considerations are relevant. First, education can be considered a process for learning how to do something. This implies that digital media may be involved in learning (as a way of doing something). When educating potential members, those from outside the community/culture may question how digital media are being used, others simply hop on board. One's ability to adapt to the use of new digital media within a community is often a key factor for achieving "learned" member status.

Given these general premises about learning we can posit that within any given digital media situation, designers should *examine* the interactional role of the new and previous digital media, *assess* the value that those digital media hold for group members, and *consider* the way that the digital media will impact interaction and the ability to form and maintain the community/culture. When considering how to provide instruction about digital media use, we advocate going beyond mere training manuals, but to consider the ways social learning about the correct way of using tools is based on cultural premises, even when the tools are new.

CULTURAL COMPARISONS

This volume has brought together seven different studies that have all been conceptualized as comparisons. That is, we are able to see what is distinct about particular practices through the juxtaposition of people who act, form relationships, and have different cultural interpretations for their actions in different scenes and situations. Even though calls have been made for HCI to consider culture (Hollan, Hutchins, and Kirsch 2000), this practice of cultural comparison is fundamentally missing from typical design processes. Cultural comparisons can be conducted at several levels, from the relational level including descriptions about the role address terms play when enacting motherhood in Columbia (Fitch 1991), or co-worker relationships are altered when using video conferencing software (Peters, chapter 3) to the cultural levels in Philipsen's (1975) description of affect displays in demonstrating how to be a proper kind of man to Poutiainen's (chapter 6) description of speaking on cell phones as an enactment of mythic Finnishness to Katriel's (1986) explanation of communicating directly as a "sabra," or mythic Jew.

Through comparisons at these different levels, we can begin to recognize both the breadth of comparisons as well as the depth of the connections between each research study. That is, when we conduct research, it adds to our store of knowledge about cultural premises, relationships and communicative acts. It would be beneficial for designers who also research specific practices to add to this store of knowledge to further our abilities to enhance interactions with one another and with the digital media that facilitate those interactions.

Designers may consider the ways these cultural identities are embedded within communicative practices, the ways they can change based on the use of new digital media, as well as the way they can produce new cultural communities of users who exhibit like practices.

LSR-INFLUENCED DESIGN

From an LSR perspective, we recognize that people share common practices, norms and values that are displayed and enacted within interactional moments. It is this set of assumptions and beliefs that we believe might serve as a guide for improving design practices. Compiling this volume based on different studies offers us all a shared glimpse of the assumptions about interaction that are embedded within a range of digital media designs.

The process of design mirrors the research process. LSR can illustrate not only the cultural premises for acting when using digital media, but also the cultural premises that designers may rely upon. By reflecting on Maguire's (2001) recommendations for designers, there are several values and beliefs about what people do, who they are to one another, and as a group that may come into play. For instance, we agree that it is good practice to involve users as key stakeholders in any design. One of the practices often advocated is to create personas. As we alluded to in the introduction and part III, personas should transcend a psychological perspective that includes personality features to focus more on the acts or interactions that create the need to display oneself as a particular kind of person. The "context of use" phase, that Maguire (2001) alludes to, is exactly what this volume has been advocating, to learn more about users' "roles, responsibilities and task goals" (599). One of the methods to get at this is the "talk aloud" practice whereby a person who is performing a task through digital media is asked to speak aloud about what he or she is doing. Having listened to recordings from this type of product-development research, noticeably not everyone is able to be as articulate during this type of activity as others. While this may be chalked up to having the right personality for this exercise, cultural norms for interaction may also be at play. For instance, as Carbaugh (1988) and others have described, even so called "sharing" behaviors are not always practiced nor valued the same everywhere.

The three main steps included in Maguire's (2001) development cycle: planning, understanding the context and requirements to produce prototypes, and the evaluation of its usability (589) are moments that were described within the model that we introduced in the introduction (see Figure 0.1). It is in the middle step, understanding local contexts, and the iteration of this process, that designers and developers are engaged in their own communicative practices with people about digital media use. Like Sprain and Boromisza-Habashi (2013), we advocate bringing LSR scholars to the table to provide just such cultural insights. Clearly, there remains much work to be done, and we are eager to begin collaborating to do it.

REFERENCES

Carbaugh, Donal. "Comments on 'Culture'; in Communication Inquiry." *Communication Reports* 1, no. 1 (1988): 38–41. doi:10.1080/08934218809367460.

Fitch, Kristine. "The Interplay of Linguistic Universals and Cultural Knowledge in Personal Address: Colombian Madre Terms." *Communication Monographs*, 58 (1991): 254–272.

Hollan, James, Edwin Hutchins, and David Kirsch. "Distributed Cognition: Toward a New Foundation for Human-Computer Interaction Research." *ACM Transactions on Computer-Human Interaction*. 7, no. 2 (2000): 174–196.

Hymes, Dell. "Toward Ethnographies of Communication." *American Anthropologist* 66, no. 6 (1964): 1–34.

Katriel, Tamar. *Talking Straight: Dugri Speech in Israeli Sabra Culture*. Cambridge, UK: Cambridge University Press, 1986.

Maguire, Martin. "Methods to Support Human-Centred Design." *International Journal of Human-Computer Studies* 55 (2001): 587–634. doi:10.1006/ijhc.2001.050.

Milburn, Trudy. "Speech Community." In *The International Encyclopedia of Language and Social Interaction,* edited by Karen Tracy, Cornelia Ilie, and Todd Sandel. Boston: John Wiley & Sons, 2015.

Philipsen, Gerry. "Speaking 'Like a Man' in Teamsterville: Culture Patterns of Role Enactment in an Urban Neighborhood." *Quarterly Journal of Speech* 61, no. 1 (1975): 13–23.

Sprain, Leah, and David Boromisza-Habashi. "The Ethnographer of Communication at the Table: Building Cultural Competence, Designing Strategic Action." *Journal of Applied Communication Research* 41, no. 2 (2013): 181–87.

Bibliography

Aakhus, Mark. "Communication as Design." *Communication Monographs* 74, no. 1 (2007): 112–117. doi:10.1080/03637750701196383.

Aakhus, Mark, and Sally Jackson. "Technology, Design, and Interaction." In *Handbook of Language and Social Interaction*, edited by Kristine L. Fitch and Robert E. Sanders. 411–36. Mahwah, NJ: Erlbaum, 2005.

Alapuro, Risto. *Suomen Synty Paikallisena Ilmiönä 1890–1933* [The Birth of Finland as a Local Phenomenon 1890–1933]. Helsinki: Tammi, 1994.

Alasuutari, Pertti. "Älymystö ja Kansakunta" [Intelligentsia and the Nation]. In *Elävänä Euroopassa. Muuttuva suomalainen identiteetti*, edited by Pertti Alasuutari and Petri Ruuska, 153–174. Tampere: Vastapaino, 1998.

Allen, I. Elaine, and Jeff Seaman. *Changing Course: Ten Years of Tracking Online Education in the United States*. Oakland, CA: Babson Survey Research Group and Quahog Research Group, 2013. February 6, 2015. http://www.onlinelearning-survey.com/reports/changingcourse.pdf.

Alred, Gerald J. "Bridging Cultures: The Academy and the Workplace." *Journal of Business Communication* 43, no. 2 (2006): 79–88. doi:10.1177/0021943605285659.

American Society for Training and Development. "Less Classroom, More Technology." *Training and Development*, May 2005.

Ang, Soon, and Linn Van Dyne. *Handbook of Cultural Intelligence: Theory, Measurement, and Applications*. Armonk, NY: M.E. Sharpe, 2008.

Anttila, Jorma. "Käsitykset Suomalaisuudesta—Traditionaalisuus ja Modernisuus" [Conceptions on Finnishness—Traditionalism and Modernness]. In *Mitä on suomalaisuus*, edited by Teppo Korhonen, 108–33. Helsinki: Suomen Antropologinen Seura, 1993.

Aoyama, Mikio. "Persona-Scenario-Goal Methodology for User-Centered Requirements Engineering." *15th IEEE International Requirements Engineering Conference*, 185–194, 2007. DOI 10.1109/RE.2007.50.

Apo, Satu. "Agraarinen suomalaisuus—rasite vai resurssi?" [Agrarian Finnishness—Burden or Resource?]. In *Olkaamme siis suomalaisia. Kalevalaseuran vuosikirja 75–76*, edited by Pekka Laaksonen and Sirkka-Liisa Mettomäki, 176–184. Helsinki: Suomalaisen Kirjallisuuden Seura, 1996.

Apo, Satu. "Suomalaisuuden stigmatisoinnin traditio" [Tradition of Stigmatizing Finnishness]. In *Elävänä Euroopassa. Muuttuva suomalainen identiteetti*, edited by Pertti Alasuutari and Petri Ruuska, 83–128. Tampere: Vastapaino, 1998.

Appel, Christine, Jackie Robbins, Joaquim More, and Tony Mullen. "Task and Tool Interface Design for L2 Speaking Interaction Online." Paper presented at the EUROCALL Conference, Gothenburg, Sweden, August 22–25, 2012.

Artino, Anthony R. "Motivational Beliefs and Perceptions of Instructional Quality: Predicting Satisfaction with Online Training." *Journal of Computer Assisted Learning* 24 (2008): 260–70. doi:10.1111/j.1365-2729.2007.00258.x.

Asmuß, Birte, and Jan Svennevig. "Meeting Talk An Introduction." *Journal of Business Communication* 46, 1 (2009): 3–22. doi:10.1177/0021943608326761.

Austin, John L. *How to Do Things with Words*. Cambridge, MA: Harvard University Press, 1962.

Bailey, Benjamin. "Communication of Respect in Interethnic Service Encounters." *Language in Society* 26, no. 3 (1997): 327–56.

Baron, Naomi S. "See You Online: Gender Issues in College Student Use of Instant Messaging." *Journal of Language and Social Psychology* 23, no. 4 (2004): 397–423.

Barrera, Isaura, and R. M. Corso. "Cultural Competency as Skilled Dialogue." *Topics in Early Childhood Special Education* 22, no. 2 (2002): 103–13. doi:10.1177/0271121402022002 0501.

Baxter, Leslie. "'Talking Things Through' and 'Putting It in Writing': Two Codes of Communication in an Academic Institution." *Journal of Applied Communication Research* 21 (1993): 313–26. doi:10.1080/00909889309365376.

Beckett, Gulbahar H., Carla Amaro-Jiménez, and Kelvin S. Beckett. "Students' Use of Asynchronous Discussions for Academic Discourse Socialization." *Distance Education* 31, no. 3 (2010): 315–35. doi:10.1080/01587919.2010.513956.

Beer, David. "The Iconic Interface and the Veneer of Simplicity: Mp3 Players and the Reconfiguration of Music Collecting and Reproduction Practices in the Digital Age." *Information, Communication & Society* 11, no. 1 (2008): 71–88.

Beldarrain, Yoany. "Distance Education Trends: Integrating New Technologies to Foster Student Interaction and Collaboration." *Distance Education* 27, no. 2 (2006): 139–153.

Ben-Yoav Nobel, Orly, Brian Wortinger, and Sean Hannah. "Winning the War and the Relationships: Preparing Military Officers for Negotiations with Non-Combatants." 2007. www.dtic.mil/cgi-bin/GetTRDoc?AD=ADA472089. U.S. Army Research Institute for the Behavioral and Social Sciences. Research Report 1877. Arlington, Virginia.

Berry, Michael. "If You Run Away from a Bear You Will Run into a Wolf: Finnish Responses to Joanna Kramer's Identity Crisis." In *Texts and Identities. Proceedings of the Third Kentucky Conference on Narratives, 1994*, edited by Joachim Knuf, 32–48. University of Kentucky: Lexington, 1995.

Berry, Michael. "The Social and Cultural Realization of Diversity: An Interview with Donal Carbaugh." *Language and Intercultural Communication* 9, no. 4 (2009): 230–241.

Blackboard, Inc. "Blackboard Exemplary Course Program Rubric." Blackboard. com. 2012. Accessed December 11, 2014. http://www.blackboard.com/

getdoc/7deaf501-4674-41b9-b2f2-554441ba099b/2012-Blackboard-Exemplary-Course-Rubric.aspx.

Blackboard, Inc. "Wikis." Blackboard Help. Accessed August 12, 2013. https://help.blackboard.com/en-us/Learn/9.1_SP_10_and_SP_11/Instructor/050_Course_Tools/Wikis#.

Blankenship, Mark. "How Social Media Can and Should Impact Higher Education." *Education Digest* 76, no. 7 (March 2011): 39–42.

Boden, Deirdre. *The Business of Talk: Organizations in Action.* Cambridge, UK: Polity Press, 1994.

Boellstorff, Tom. *Coming of Age in Second Life: An Anthropologist Explores the Virtually Human.* Princeton, NJ: University Press, 2008.

Bolden, Galina. "Reopening Russian Conversations: The Discourse Particle *-to* and the Negotiation of Interpersonal Accountability in Closings." *Human Communication Research* 34 (2008): 99–136.

Bormann, Ernest G. "Symbolic Convergence: Organizational Communication and Culture." In *Communication and Organizations, an Interpretive Approach,* edited by Linda Putnam and Michael E. Pacanowsky, 100. Beverly Hills: Sage Publications, 1983.

Boromisza-Habashi, David, and Russell M. Parks. "The Communal Function of Social Interaction on an Online Academic Newsgroup." *Western Journal of Communication* 78, no. 2 (2014): 194–212.

Botturi, Luca. "Design Models as Emergent Features: An Empirical Study in Communication and Shared Mental Models in Instructional Design." *Canadian Journal of Learning and Technology* 32, no. 2 (Spring 2006). Accessed February 6, 2015. http://www.cjlt.ca/index.php/cjlt/article/view/50/47.

Bouwmeester, Maaike. "Examining the Effects of Reflective Rubrics in the E-portfolio Peer Review Process on Pre-service Teachers' Ability to Integrate Academic Coursework and Field Experiences." PhD dissertation, New York University, 2011. ProQuest (UMI 3478271).

Bradford, Peter, Margaret Porciello, Nancy Balkon, and Debra Backus. "The Blackboard Learning System: The Be All and End All in Educational Instruction?" *Journal of Educational Technology Systems* 35, no. 3 (2007): 301–14. doi:10.2190/X137-X73L-5261-5656.

Branzburg, Jeffrey, and Kristen Kennedy. "Online Professional Development." *Technology and Learning* 22, no. 2 (2001): 18–27.

Brunk-Chavez, Beth L., and Shawn J. Miller. "Decentered, Disconnected, and Digitized: The Importance of Shared Space." *Kairos* 11, no. 1 (Spring 2007). Accessed December 21, 2014. http://kairos.technorhetoric.net/11.2/topoi/brunk-miller/decenteredtest/decentered.pdf.

Bryant, Susan. "The Five Habits: Building Cross-cultural Competence in Lawyers." *Clinical Law Review,* Fall 2001. Accessed February 6, 2015. http://www.law.cuny.edu/faculty/pedagogy/bryant-habits.pdf.

Bryce, Nadine. "Teacher Candidates' Collaboration and Identity in Online Discussions." *Journal of University Teaching and Learning Practice* 11, no. 1 (2014). Accessed February 6, 2015. http://ro.uow.edu.au/jutlp/vol11/iss1/7/.

Buchanan, Richard. "Wicked Problems in Design Thinking." *Design Issues,* 8, no. 2 (1992): 5–21.

Burns, Louise, Meredith Marra, and Janet Holmes. "Women's Humour in the Workplace: A Quantitative Analysis." *Australian Journal of Communication* 28 1 (2001): 83–108. Accessed February 6, 2015. http://search.informit.com.au/doc umentSummary;dn=200112196;res=IELAPA.

Caddick, Richard, and Steve Cable. *Communicating the User Experience: A Practical Guide for Creating Useful UX Documentation.* West Sussez, UK: Wiley & Sons, 2011.

Cameron, Deborah. *Good to Talk? Living and Working in a Communication Culture.* London: Sage, 2000a.

Cameron, Deborah. "Styling the Worker: Gender and the Commodification of Language in the Globalized Service Economy." *Journal of Sociolinguistics* 4, no. 3 (2000b): 323–47.

Cameron, Deborah. "Talk from the Top Down." *Language & Communication* 28 (2008): 143–55.

Campion, Michael C., Robert E. Ployhart, and William I. MacKenzie, Jr. "The State of Research on Situational Judgment Tests: A Content Analysis and Directions for Future Research." *Human Performance* 27, no. 4 (2014). doi:10.1080/0895928 5.2014.929693.

Canaves, Sky. "China's Social Networking Problem." *Spectrum, IEEE* 48, no. 6 (June 2011): 74–77.

Carbaugh, Donal. *Talking American: Cultural Discourses on Donahue.* Norwood, NJ: Ablex, 1988a.

Carbaugh, Donal. "Comments on 'Culture'; in Communication Inquiry." *Communication Reports* 1, no. 1 (1988b): 38–41. doi:10.1080/08934218809367460.

Carbaugh, Donal. "Toward a Perspective on Cultural Communication and Intercultural Contact." *Semiotica* 80, no. 1–2 (1990): 15–36. doi:10.1515/semi.1990.80.1-2.15.

Carbaugh, Donal. "Personhood, Positioning and Cultural Pragmatics: American Dignity in Cross-Cultural Perspective." In *Communication Yearbook*, volume 17, edited by Stanley A. Deetz, 159–186. New York, NY: Routledge, 1994.

Carbaugh, Donal. "The Ethnographic Communication Theory of Philipsen and Associates." In *Watershed Research Traditions in Human Communication Theory*, edited by Donald P. Cushman and Branislav Kovacic. Albany, NY: State University of Albany Press, 1995.

Carbaugh, Donal. *Situating Selves: The Communication of Social Identities in American Scenes.* Albany, NY: SUNY Press, 1996.

Carbaugh, Donal, *Cultures in Conversation.* Mahwah, NJ: Lawrence Erlbaum Associates, Inc., 2005.

Carbaugh, Donal. "Cultural Discourse Analysis: Communication Practices and Intercultural Encounters." *Journal of Intercultural Communication Research* 36, no. 3 (2007): 167–182. doi:10.1080/17475750701737090.

Carbaugh, Donal. "Ethnography of Communication." In *The International Encyclopedia of Communication*, edited by Wolfgang Donsbach. Hoboken, NJ: Wiley-Blackwell, 2008.

Carbaugh, Donal. "A Communication Theory of Culture." In *Inter/Cultural Communication: Representation and Construction of Culture*, edited by Anastacia Kurylo, 69–87. Thousand Oaks, CA: Sage, 2012.

Carbaugh, Donal, Timothy Gibson, and Trudy Milburn. "A View of Communication and Culture: Scenes in an Ethnic Cultural Center and a Private College." In *Emerging Theories of Human Communication*, edited by Branislav Kovacic, 1–24. Albany, NY: SUNY Press, 1997.

Carbaugh, Donal, and Sally O. Hastings. "A Role for Communication Theory in Ethnography and Cultural Analysis." *Communication Theory* 2, no. 2 (1992): 156–65. doi:10.1111/j.1468-2885.1992.tb00035.x.

Carbaugh, Donal, and Sunny Lie. "Competence in Interaction: Cultural Discourse Analysis." In *Intercultural Communication Competence: Conceptualization and Its Development in Cultural Contexts and Interactions*, edited by Xiaodong Dai and Guo-Ming Chen, 69–81. Cambridge: Cambridge Scholars Publishing, 2014.

Carbaugh, Donal, Elizabeth Molina-Markham, Brion van Over, and Ute Winter. "Using Communication Research for Cultural Variability in Human Factor Design." In *Advances in Human Aspects of Road and Rail Transportation*, edited by Neville Stanton, 176–185. Boca Raton, FL: CRC Press, 2012.

Carbaugh, Donal, Elena V. Nuciforo, Makoto Saito, and Dong-Shin Shin. "'Dialogue' in Cross-Cultural Perspective: Japanese, Korean, and Russian Discourses." *Journal of International and Intercultural Communication* 4, no. 2 (2011): 87–108. doi:10.1080/17513057.2011.557500.

Carbaugh, Donal, Ute Winter, Elizabeth Molina-Markham, Brion van Over, Sunny Lie, and Timothy Grost. "A Model for Investigating Cultural Dimensions of Communication in the Car." Manuscript submitted for publication, 2014.

Carbaugh, Donal, Ute Winter, Brion van Over, Elizabeth Molina-Markham, and Sunny Lie. "Cultural Analyses of in-Car Communication." *Journal of Applied Communication Research* 41, no. 2 (2013): 195–201.

Carroll, John M., "Human Computer Interaction—Brief Introduction." In *The Encyclopedia of Human-Computer Interaction*, 2nd edition, edited by Mads Soegaard and Rikke Friis Dam. Aarhus, Denmark: The Interaction Design Foundation, 2013. Accessed February 6, 2015. http://www.interaction-design.org/encyclopedia/human_computer_interaction_hci.html.

Center for Instructional & Learning Technologies. Blackboard Tips for Faculty. Accessed June 16, 2010. http://und.edu/academics/cilt/instructional-design/weekly-tips/weeklytipsjune16.pdf.

Chandler, Jennifer V. "Why Culture Matters: An Empirically-based Pre-deployment Training Program." Master's thesis, Naval Postgraduate School, 2005. Accessed February 6, 2015. http://www.dtic.mil/dtic/tr/fulltext/u2/a439382.pdf.

Chang, Hui-Ching. *Clever, Creative, Modest: The Chinese Language Practice*. Shanghai, China: Shanghai Foreign Language Education Press, 2010.

Chen, Yongdong. "Wechat and Weibo: Complement Not Substitutes [In Chinese]." *News and Writing* 4 (2013): 31–33.

Chin, Yann-Ling. "'Platonic Relationships' in China's Online Social Milieu: A Lubricant for Banal Everyday Life." *Chinese Journal of Communication* 4, no. 4 (2011): 400–416.

Christian, Michael S., Bryan D. Edwards, and Jill C. Bradley. "Situational Judgment Tests: Constructs Assessed and a Meta-Analysis Of Their Criterion-

Related Validities." *Personnel Psychology* 63, no. 1 (2010): 83–117. doi:10.1111/j.1744-6570.2009.01163.x.

Cleaver, Samantha. "Beyond Blackboard and into Virtual Communities." *Diverse: Issues in Higher Education* 25, no. 18 (October 2008): 32. Accessed December 27, 2014. ProQuest.

Cooper, Alan. *The Inmates Are Running the Asylum: Why High Tech Products Drive Us Crazy and How to Restore the Sanity.* Indianapolis, IN: Sams Publishing, 2004.

Coupland, Nikolas, and Justine Coupland. "Language, Ageing and Ageism." In *The New Handbook of Language and Social Psychology*, edited by Peter Robinson and Howard Giles, 465–486. New York: John Wiley, 2001.

Coutu, Lisa. "Communication Codes of Rationality and Spirituality in the Discourse of and about Robert S. Mcnamara's 'in Retrospect.'" *Research on Language and Social Interaction* 33, no. 2 (2000): 179–211.

Coutu, Lisa. "Contested Social Identity and Communication in Talk and Text About the Vietnam War." *Research on Language & Social Interaction* 41, no. 4 (2008): 387–407.

Cox, Taylor. *Cultural Diversity in Organizations: Theory, Research and Practice.* San Francisco: Berrett-Koehler, 1994.

Crandall, Sonia J., Gavin George, Gail S. Marion, and Stephen Davis. "Applying Theory to the Design of Cultural Competency Training for Medical Students: A Case Study." *Academic Medicine* 78, no. 6 (2003): 588–94.

Crosson, Jesse C., Weiling Deng, Chantal Brazeau, Linda Boyd, and Maria Soto-Greene. "Evaluating the Effect of Cultural Competency Training on Medical Student Attitudes." *Family Medicine* 36, no. 3 (March 2004): 199–203. Accessed February 6, 2015. http://www.stfm.org/fmhub/fm2004/March/Jesse199.pdf.

Dantin, Ursula. "Application of Personas in User Interface Design for Educational Software." In *Australasian Computing Education Conference. Conferences in Research and Practice in Information Technology*, edited by A. Young and D. Tolhurst, 42, 239–247. Newcastle, Australia. 2005.

de Souza e Silva, Adriana. "From Cyber to Hybrid: Mobile Technologies as Interfaces of Hybrid Spaces." *Space and Culture* 9, no. 3 (2006): 261–78.

Deetz, Stanley, ed. *Communication Yearbook*, volume 17. New York: Routledge, 1994.

Dix, Alan, Janet E. Finlay, Gregory D. Abowd, and Russell Beale. *Human-Computer Interaction*, 3rd edition. Essex, England: Pearson Education Limited, 2004.

Dogancay, Seran. "Your Eye is Sparkling: Formulaic Expressions and Routines in Turkish." *Working Papers in Educational Linguistics* 6, no. 2 (1990): 49–64.

Dori-Hacohen, Gonen, and Nimrod Shavit. "The Cultural Meanings of Israeli Tokbek (Talk-Back Online Commenting) and Their Relevance to the Online Democratic Public Sphere." *International Journal of Electronic Governance* 6, no. 4 (2013): 361–79.

Dringus, Laurie P., and Timothy Ellis. "Temporal Transitions in Participation Flow in an Asynchronous Discussion Forum." *Computers & Education* 54, no. 2 (2009): 340–49. doi:10.1016/j.compedu.2009.08.011.

DuFon, Margaret. "The Socialization of Leave-Taking in Indonesian." *Pragmatics and Language Learning* 12 (2010): 91–111.

Earley, Christopher P., and Soon Ang. *Cultural Intelligence: Individual Interactions across Cultures.* Stanford, CA: Stanford University Press, 2003.

Edgerly, Louisa. "Difference and Political Legitimacy: Speakers' Construction of 'Citizen' and 'Refugee' Personae in Talk About Hurricane Katrina." *Western Journal of Communication* 75, no. 3 (2011): 304–22.

Edwards, Derek, and Jonathan Potter. *Discursive Psychology*. Thousand Oaks, CA: Sage, 1992.

Ellis, Donald G., and B. Aubry Fisher. *Small Group Decision Making: Communication and the Group Process*, 4th Edition. New York: McGraw-Hill, Inc., 1994.

Emerson, Robert M., Rachel I. Fretz, and Linda L. Shaw. *Writing Ethnographic Fieldnotes*. Chicago, IL: The University of Chicago Press, 1995.

Fenton, Celeste, and Brenda W. Watkins. "Online Professional Development for K–12 Educators: Benefits for School Districts with Applications for Community College Faculty Professional Development." *Community College Journal of Research and Practice* 31, no. 6 (2007): 531–33. doi:10.1080/10668920701359946.

Finn, Kathleen E. "Introduction: An Overview of Video-mediated Communication Literature." In *Video-mediated Communication*, edited by Kathleen E. Finn, Abigail J. Sellen, and Sylvia B. Wilbur, 3–21. Mahwah, NJ: Lawrence Erlbaum Associates, 1997.

Firth, Raymond. "Verbal and Bodily Rituals of Greeting and Parting." In *The Interpretation of Ritual*, edited by J. S. La Fontaine, 1–38. London: Tavistock, 1972.

Fitch, Kristine. "A Ritual for Attempting Leave-Taking in Colombia." *Research on Language & Social Interaction* 24, no. 1–4 (1990): 209–224.

Fitch, Kristine. "The Interplay of Linguistic Universals and Cultural Knowledge in Personal Address: Colombian Madre Terms." *Communication Monographs* 58 (1991): 254–272.

Fitch, Kristine. "Culture, Ideology and Interpersonal Communication Research." In *Communication Yearbook*, volume 17, edited by Stanley A. Deetz, 104–135. New York: Routledge, 1994.

Fitch, Kristine. *Speaking Relationally: Culture, Communication, and Interpersonal Connection*. New York: The Guilford Press, 1998.

Fitch, Kristine L. "Pillow Talk?" *Research on Language and Social Interaction* 32, no. 1–2 (1999): 41–50.

Flyvbjerg, Bent. *Making Social Science Matter: Why Social Inquiry Fails and How It Can Succeed Again*. Oxford, UK: Cambridge University Press, 2001.

Fong, Mary. "'Luck Talk' in Celebrating the Chinese New Year." *Journal of Pragmatics* 32 (2000): 219–37.

Fortunati, Leopoldina. "Italy: Stereotypes, True and False." In *Perpetual Contact: Mobile Communication, Private Talk, Public Performance*, edited by James Katz and Mark Aakhus, 42–62. Cambridge, MA: Cambridge University Press, 2002.

Friedman, Batya, ed. *Human Values and the Design of Computer Technology*. New York: Cambridge University Press, 1997.

Gane, Nicholas, and David Beer. *New Media: The Key Concepts*. Oxford: Berg, 2008.

Gao, Ge, and Stella Ting-Toomey. *Communicating Effectively with the Chinese*. Thousand Oaks, CA: Sage, 1998.

Garfinkel, Harold. "Studies of the Routine Grounds of Everyday Activities." *Social Problems* 11, no. 3 (1964): 225–250.

Geertz, Clifford. "Deep Play: Notes on the Balinese Cockfight." *Daedalus* 101, no. 1 (1972): 1–37.

Gioia, Dennis A., and Peter P. Poole. "Scripts in Organizational Behavior." *Academy of Management Review* 9, no. 3 (1984): 449–59.

Goffman, Erving. *Relations in Public: Microstudies of the Social Order*. Philadelphia, PA: Harper and Row, 1971.

Goffman, Erving. *Frame Analysis: An Essay on the Organization of Experience*. Boston, MA: Northeastern University Press, 1974.

Gordon, Tuula, Katri Komulainen, and Kirsti Lempiäinen. *"Suomineitonen hei! Kansallisuuden sukupuoli"* [Greeting Young Girl Finland! Gender of Nationality]. Tampere: Vastapaino, 2002.

Grbich, Carol. *Qualitative Data Analysis: An Introduction*. London: SAGE Publications, 2007.

Greene-Sands, Robert, and Allison Greene-Sands, eds. *Cross-Cultural Competence for a 21st Century Military*. Lanham, MD: Lexington, 2014.

Guan, Xin, and Todd L. Sandel. "The Acculturation and Identity of New Immigrant Youth in Macao." *China Media Research*, in press.

Gustafson, Kent L., and Robert M. Branch. *Survey of Instructional Development Models*, 4th edition. 2002.

Haddington, Pentti. "Turn-Taking for Turntaking: Mobility, Time, and Action in the Sequential Organization of Junction Negotiations in Cars." *Research on Language and Social Interaction* 43, no. 4 (2010): 372–400.

Häikiö, Martti. "Kännykät ja suomalainen kansanluonne" [Mobile Phones and the Finnish Folk Psychology]. *Kanava* 9 (1998): 571–2.

Hao, Zhidong. *Macau: History and Society*. Hong Kong: Hong Kong University Press, 2011.

Hara, Noriko, Curtis J. Bonk, and Charoula Angeli. "Content Analysis of Online Discussion in an Applied Educational Psychology Course." *Instructional Science* 28, no. 2 (2000): 115–52. doi:10.1023/A:1003764722829.

Harrison, Theresa. "Communication and Interdependence in Democratic Organizations." In *Communication Yearbook*, volume 17, edited by Stanley A. Deetz, 247–274. New York: Routledge, 1994.

Hart, Tabitha. "Speech Codes Theory as a Framework for Analyzing Communication in Online Educational Settings." In *Computer Mediated Communication: Issues and Approaches in Education*, edited by Sigrid Kelsey and Kirk St. Amant. Hershey, PA: IGI Global, 2011.

Hart, Tabitha. "Learning How to Speak Like a 'Native': A Case Study of a Technology-Mediated Oral Communication Training Program." *Journal of Business and Technical Communication* (2016).

Hartford, Beverly S., and Kathleen Bardovi-Harlig. "Closing the Conversation: Evidence from the Academic Advising Session." *Discourse Processes* 15, no. 1 (1992): 93–116.

Hew, Khe F., and Wing Sum Cheung. "Students' Use of Asynchronous Voice Discussion in a Blended Learning Environment: A Study of Two Undergraduate Classes." *Electronic Journal of E-Learning* 10, no. 4 (2012): 360–67. www.ejel.org/issue/download.html?idArticle=215.

Hew, Khe Foon, Wing Sum Cheung, and Connie Siew Ling Ng. "Student Contribution in Asynchronous Online Discussion: A Review of the Research and Empirical Exploration." *Instructional Science* 38, no. 6 (2010): 571–606. doi:10.1007/s11251-008-9087-0.

Ho, Melissa R., Thomas N. Smyth, Matthew Kam, and Deardon, Andy. "Human-Computer Interaction for Development: The Past, Present, and Future." *ITI: Information Technologies and International Development* 5, no. 4 (2009): 1–18.

Holdford, David. "Service Scripts: A Tool for Teaching Pharmacy Students How to Handle Common Practice Situations." *American Journal of Pharmaceutical Education* 70, no. 1 (2006): 1–7.

Hollan, James, Edwin Hutchins, and David Kirsch. "Distributed Cognition: Toward a New Foundation for Human-Computer Interaction Research." *ACM Transactions on Computer-Human Interaction* 7, no. 2 (2000): 174–196.

Holmes, Janet. "Politeness, Power, and Provocation: How Humor Functions." *Discourse Studies* 2 (2000): 159–185. doi:10.1177/1461445600002002002.

Holmes, Janet. *Gendered Talk at Work: Constructing Gender Identity through Workplace Discourse.* Malden, MA: Blackwell, 2008.

Holsombach-Ebner, Cinda. "Quality Assurance in Large Scale Online Course Production." *Online Journal of Distance Learning Administration* 16, no. 2 (September 1, 2013). Accessed February 6, 2015. http://www.westga.edu/~distance/ojdla/fall163/holsombach-ebner164.html.

Holzer, Elie. "Professional Development of Teacher Educators in Asynchronous Electronic Environment: Challenges, Opportunities and Preliminary Insights from Practice." *Educational Media International* 41, no. 1 (2004): 81–89. doi:10.1080/0952398032000105139.

Homsey, Dini, and Todd L. Sandel. "The Code of Food and Tradition: Exploring a Lebanese (American) Speech Code in Practice in Flatland." *Journal of Intercultural Communication Research* 41, no. 1 (2012): 59–80.

Hong, Kaylene. "WeChat Climbs to 438 Million Monthly Active Users." *The Next Web.* August 13, 2014. Accessed January 14, 2015. http://thenextweb.com/apps/2014/08/13/wechat-climbs-to-438-million-monthly-active-users-closing-in-on-whatsapps-500-million/.

Hou, E. "Government Wechat, Expired 'Old Ticket' [In Chinese]." *China Media Report Overseas* 10, no. 1 (2014): 1–7.

Huspek, Michael. "Oppositional Codes and Social Class Relations." *The British Journal of Sociology* 45, no. 1 (1994): 79–102.

Hymes, Dell. "The Ethnography of Speaking." In *Anthropology and Human Behavior*, edited by Thomas Gladwin and William Sturtevant, 13–53. Washington, DC: The Anthropological Society of Washington, 1962.

Hymes, Dell. "Toward Ethnographies of Communication." *American Anthropologist* 66, no. 6 (1964): 1–34.

Hymes, Dell. "Models for the Interaction of Language and Social Life." In *Directions in Sociolinguistics: The Ethnography of Communication*, edited by John Gumperz and Dell Hymes, 35–71. New York: Basil Blackwell Inc., 1972a.

Hymes, Dell. "On Communicative Competence." In *Sociolinguistics*, edited by J. B. Pride and Janet Holmes, 269–85. Baltimore, MD: Penguin Education, 1972b.

Hymes, Dell. *Foundations in Sociolinguistics: An Ethnographic Approach.* Philadelphia, PA: University of Pennsylvania Press, 1974.

Jackson, Michele H., Marshall S. Poole, and Timothy Kuhn. "The Social Construction of Technology in Studies of the Workplace." In *Handbook of New Media*, edited by Leah A. Lievrouw and Sonia Livingstone, 236–253. Thousand Oaks, CA: Sage, 2002.

Jackson, Sally, and Mark Aakhus. "Becoming More Reflective About the Role of Design in Communication." *Journal of Applied Communication Research* (2014): 1–10.

Jeffreys, Marianne. "Dynamics of Diversity: Becoming Better Nurses through Diversity Awareness." *NSNA Imprint*, November/December 2008, 36–41. Accessed February 6, 2015. http://www.nsna.org/Portals/0/Skins/NSNA/pdf/Imprint_NovDec08_Feat_Jeffreys.pdf.

Katriel, Tamar. *Talking Straight: Dugri Speech in Israeli Sabra Culture*. Cambridge, UK: Cambridge University Press, 1986.

Katriel, Tamar, and Gerry Philipsen. "'What We Need Is Communication': 'Communication' as a Cultural Category in Some American Speech." *Communications Monographs* 48 (1981): 301–17. doi:10.1080/03637758109376064.

Katzenstein, Mary F., and Judith Reppy. *Beyond Zero Tolerance: Discrimination in Military Culture*. Lanham, MD: Rowman and Littlefield Publishers, 1999.

Kay, Alan. "User Interface: A Personal View." In *The Art of Human-Computer Interface Design*, edited by Brenda Laurel and S. Joy Mountford, 191–207. Reading, MA: Addison-Wesley, 1990.

Keating, Elizabeth. "The Ethnography of Communication." In *Handbook of Ethnography*, edited by Paul Atkinson, Amanda Coffey, Sara Delamont, John Lofland, and Lyn Lofland, 285–300. London: Sage, 2001.

Keengwe, Jared, and Jung-Jin Kang. "Blended Learning in Teacher Preparation Programs: A Literature Review." *International Journal of Information and Communication Technology Education* 8 (2012).

Kelly, Rob. "Building Community in Self-Paced Online Courses." *Online Cl@ssroom* 1 (June 2005): 7. Accessed February 6, 2015. http://www.magnapubs.com/newsletter/online-classroom/16/building_community_in_self_paced_online_courses-10211-1.html.

Kendall, Lori. *Hanging out in the Virtual Pub: Masculinities and Relationships Online*. Berkeley: University of California Press, 2002.

Kenny, Richard F., Zuochen Zhang, Richard A. Schwier, and Katy Campbell. "A Review of What Instructional Designers Do: Questions Answered and Questions Not Asked." *Canadian Journal of Learning and Technology* 31, no. 1 (Winter 2005): 9–16. Accessed February 6, 2015. http://www.cjlt.ca/index.php/cjlt/article/view/147/140.

Keyton, Joanne. "Relational Communication in Groups." In *The Handbook of Group Communication Theory and Research*, edited by Larry R. Frey, Dennis S. Gouran, and Marshall Scott Poole, 192–222. Thousand Oaks, CA: Sage, 1999.

Khine, Myint Swe, and Atputhasamy Lourdusamy. "Blended Learning Approach in Teacher Education: Combining Face-to-Face Instruction, Multimedia Viewing and Online Discussion." *British Journal of Educational Technology* 34 (2003): 671–675.

Kimbell, Lucy. "Rethinking Design Thinking: Part 1." *Design and Culture* 3 (2011): 285–306.

Kimbell, Lucy. "Rethinking Design Thinking: Part 2." *Design and Culture* 4 (2012): 129–148.

Knapp, Mark L., Roderick P. Hart, Gustav W. Friedrich, and Gary M. Shulman. "The Rhetoric of Goodbye: Verbal and Nonverbal Correlates of Human Leave-Taking." *Communication Monographs* 40, no. 3 (1973): 182–198.

Kohvakka, Rauli. "Yhteisöpalvelut istuvat suomalaiseen sosiaalisuuteen" [Social Network Services Fit into Finnish Sociality]. *Hyvintointikatsaus* 2 (2013): 58–62. Helsinki, Finland: Tilastokeskus.

Korthagen, Fred A. *Linking Practice and Theory*. New Jersey: Lawrence Erlbaum Associates, 2001.

Krumm, Stefan, Filip Lievens, Joachim Hüffmeier, Anastasiya A. Lipnevich, Hanna Bendels, and Guido Hertel. "How 'Situational' is Judgment in Situational Judgment Tests?" *Journal of Applied Psychology* (August 11, 2014): 1–18. doi:10.1037/a0037674.

Kuhlenschmidt, Sally. "Why Use Blogs and Wikis in Blackboard?" Lecture, January 21, 2009. Accessed February 6, 2015. https://www.wku.edu/teaching/online/blackboardblogswikis.ppt.

Lackey, Dundee. "Why Wiki?" *Kairos Praxis Wiki* 12, no. 1 (2007). Accessed February 6, 2015. http://praxis.technorhetoric.net/tiki-index.php?page=Why_Wiki.

Laird, Phillip G. "Integrated Solutions to ELearning Implementation: Models, Structures and Practices at Trinity Western University." *Online Journal of Distance Learning Administration* 7, no. 3 (2004). Accessed February 6, 2015. http://www.westga.edu/~distance/ojdla/fall73/laird73.html.

Larson, David K., and Chung-Hsien Sun. "Comparing Student Performance: Online Versus Blended Versus Face-to-Face." *Journal of Asynchronous Learning Networks* 13, no. 1 (2009): 31–42. Accessed February 6, 2015. https://idt7895.files.wordpress.com/2009/05/comparing-student-performance-in-different-delivery-methods.pdf.

Laurier, Eric, Barry Brown, and Hayden Lorimer. *Habitable Cars: The Organisation of Collective Private Transport*. Full Research Report to the Economic Social Research Council (University of Edinburgh, University of Glasgow), RES-000–23–0758. Swindon: ESRC, 2007.

Laurier, Eric, Hayden Lorimer, Barry Brown, Owain Jones, Oskar Juhlin, Allyson Noble, Mark Perry, Daniele Pica, Philippe Sormani, Ignaz Strebel, Laurel Swan, Alex S. Taylor, Laura Watts, and Alexandra Weilenmann. "Driving and 'Passengering': Notes on the Ordinary Organization of Car Travel." *Mobilities* 3, no. 1 (2008): 1–23.

Lave, Jean, and Etienne Wenger. *Situated Learning: Legitimate Peripheral Participation*. Cambridge: Cambridge University Press, 1991.

Lee, Hung-Tien. "Songshan xian tong che, ditou zu ye kandedao!: Songjiang Nanjing zhan she da xing yin dao di tie" [Songshan line, even the head lowered tribe can see!: Songjiang Nanjing station has set up a very large passenger direction sign]. *NOWnews* 11 (2014): 11. Accessed November 12, 2014. http://www.nownews.com/n/2014/11/11/1500924.

Lee, Silvia W., and Chin-Chung Tsai. "Identifying Patterns of Collaborative Knowledge Exploration in Online Asynchronous Discussions." *Instructional Science* 39, no. 3 (May 2011): 321–47. doi:10.1007/s11251-010-9131-8.

Lehtonen, Mikko, Olli Löytty, and Petri Ruuska. *Suomi toisin sanoen* [Putting Finland in Other Words]. Tampere: Vastapaino, 2004.

Leighter, James L., and Laura Black. "'I'm Just Raising the Question': Terms for Talk and Practical Metadiscursive Argument in Public Meetings." *Western Journal of Communication* 74, no. 5 (2010): 547–69.

Leighter, James L., Lisa Rudnick, and Theresa J. Edmonds. "How the Ethnography of Communication Provides Resources for Design." *Journal of Applied Communication Research* 41, no. 2 (2013): 209–15.

Lemis, Dafna, and Akiba A. Cohen. "Tell Me about Your Mobile and I'll Tell You Who You Are: Israelis Talk about Themselves." In *Mobile Communication: Re-Negotiation of the Social Sphere*, edited by Richard Ling and Per E. Pedersen, 187–202. London: Springer, 2005.

Leskinen, Tatu. "Kansa, kansallisuus ja sivistys" [Nation, Nationality and Education]. *Jargonia* 6, no. 3 (2005). Accessed January 20, 2015. https://jyx.jyu.fi/dspace/handle/123456789/20064.

Leung, Kwok, Soon Ang, and Mei Ling Tan. "Intercultural Competence." *Annual Review of Organizational Psychology and Organizational Behavior* 1, no. 1 (2014): 489–519. doi:10.1146/annurev-orgpsych-031413-091229.

Levin, Barbara B., Ye He, and Holly H. Robbins. "Comparative Analysis of Preservice Teachers' Reflective Thinking in Synchronous Versus Asynchronous Online Case Discussions." *Journal of Technology and Teacher Education* 14, no. 3 (2006): 439–60. libres.uncg.edu/ir/uncg/f/B_Levin_Comparative_2006.pdf.

Li, Zhuo. *Weixin dui qingnian qunti ren ji guanxi yingxiang de yanjiu [The study of WeChat's influence on young people's relationships]*. Master's thesis. Inner Mongolia University, 2014.

Licona, Miguel M., and Binod Gurung. "Asynchronous Discussions in Online Multicultural Education." *Multicultural Education* (Fall 2011): 2–8. Accessed February 6, 2015. http://files.eric.ed.gov/fulltext/EJ986884.pdf.

Lin, Hong. "Blending Online Components into Traditional Instruction in Pre-Service Teacher Education: The Good, the Bad, and the Ugly." *International Journal for the Scholarship of Teaching and Learning* 2, no. 1 (2008): 1–14.

Lin, Qiuyun. "Student Satisfactions in Four Mixed Courses in Elementary Teacher Education Program." *Internet and Higher Education* 11, no. 1 (2008): 53–59. Accessed February 6, 2015. http://eric.ed.gov/?id=EJ796651.

Lindlof, Thomas R., and Brian C. Taylor. *Qualitative Communication Research Methods*, 2nd edition. Thousand Oaks, CA: Sage, 2002.

Lock, Jennifer V. "A New Image: Online Communities to Facilitate Teacher Professional Development." *Journal of Technology and Teacher Education* 14 (2006): 663–678.

Lord, Gillian, and Lara L. Lomicka. "Developing Collaborative Cyber Communities to Prepare Tomorrow's Teachers." *Foreign Language Annals* 37, no. 3 (2004): 401–08. doi:10.1111/j.1944-9720.2004.tb02698.x.

Lorenzetti, Jennifer P. "Lessons Learned about Student Issues in Online Learning." *Distance Education Report* 9, no. 18 (2005): 1–7.

Lortie, Dan. *Schoolteacher: A Sociological Study*. Chicago: University of Chicago Press, 1975.

Mackenzie, Lauren. "Strategic Enablers: How Intercultural Communication Skills Advance Micro-Level International Security." *Journal of Culture, Language & International Security* 1, no. 1 (Summer 2014). Accessed February 6, 2015. http://culture.af.mil/library/pdf/MackenzieJCLISMay2014.pdf.

Mackenzie, Lauren, Patricia Fogerty, and Angelle Khachadoorian. "A Model for Military Online Culture Education: Key Findings and Best Practices." *EDUCAUSE Review* 48, no. 4 (July/August 2013). Accessed February 6, 2015. http://www.educause.edu/ero/article/model-online-military-culture-education-key-findings-and-best-practices.

Mackenzie, Lauren, and Megan Wallace. "Distance Learning Designed for the U.S. Air Force." *Academic Exchange Quarterly* 16, no. 2 (Summer 2012): 55–60.

Mackenzie, Lauren, and Megan Wallace. "Cross-Cultural Communication Contributions to Professional Military Education: A Distance Learning Case Study." In *Cross-Cultural Competence for a 21st Century Military: Culture, the Flipside of COIN*, edited by Robert R. Greene-Sands and Allison Greene-Sands, 239–58. Lanham, MD: Lexington Books, 2014.

Mäenpää, Pasi. "Tietoyhteiskunta ja uusi suomalaisuus" [Information Society and New Finnishness]. In *Suomen kulttuurihistoria, osa 4. Koti, kylä, kaupunki*, edited by Kirsi Saarikangas and Minna Sarantola-Weiss, 517–8. Helsinki: Tammi, 2004.

Maguire, Martin. "Methods to Support Human-Centred Design." *International Journal of Human-Computer Studies* 55 (2001): 587–634. doi:10.1006/ijhc.2001.050.

Manning, Michelle. "Building Presence and Community in the Online Class." *Kairos: Rhetoric, Technology, and Pedagogy* (February 27, 2007). Accessed February 6, 2015. http://praxis.technorhetoric.net/tiki-index.php?page=Building_Presence.

Manovich, Lev. *The Language of New Media*. Cambridge, MA: MIT Press, 2001.

Manovich, Lev. "New Media from Borges to Html." In *The New Media Reader*, edited by Noah Wardrip-Fruin and Nick Montfort. Cambridge, MA: MIT Press, 2003.

Mansvelder-Longayroux, Désirée Danièle, Douwe Beijaard, and Nico Verloop. "The Portfolio as a Tool for Stimulating Reflection of Student Teachers." *Teaching and Teacher Education* 23 (2007): 47–62.

Margolin, Victor, and Richard Buchanan, eds. *The Idea of Design: A Design Issues Reader*. Cambridge, MA: MIT Press, 1995.

May, Larissa, Kimberly D. Acquaviva, Annette Dorfman, and Laurie Posey. "Medical Student Perceptions of Self-Paced, Web-Based Electives: A Descriptive Study." *American Journal of Distance Education* 23, no. 4 (2009): 212–23. doi:10.1080/08923640903332120.

McDaniel, Michael A., Deborah L. Whetzel, and Nhung T. Nguyen. *Situational Judgment Tests for Personnel Selection*. In *An IPMAAC Personel Assessment Monograph*, edited by Leilani Yan. Alexandria, VA: IPMA-HR Assessment Council, 2006.

McKay, Everett N. *UI is Communication: How to Design Intuitive, User Centered Interfaces by Focusing on Effective Communication*. Waltham, MA: Elsevier, 2013.

Milburn, Trudy. "Speech Community: Reflections upon Communication." In *Communication Yearbook*, volume 28, edited by Pamela J. Kalbfleisch, 411–441, Lawrence Erlbaum/ICA, 2004.

Milburn, Trudy. *Nonprofit Organizations: Creating Membership through Communication*. Cresskill, NJ: Hampton Press, 2009.

Milburn, Trudy. "S.P.E.A.K.I.N.G.: A Research Tool." *Communication Institute for Online Scholarship (CIOS)*. Accessed January 1, 2015. http://www.cios.org/encyclopedia/ethnography/4speaking.htm.

Milburn, Trudy. "Speech Community." In *The International Encyclopedia of Language and Social Interaction*, edited by Karen Tracy, Cornelia Ilie, and Todd Sandel. Boston: John Wiley & Sons, 2015.

Miller, Derick, and Lisa Rudnick. *The Security Needs Assessment Protocol: Improving Operational Effectiveness through Community Security*. Report. New York and Geneva: United Nations Publications, 2008.

Miller, Derick, and Lisa Rudnick. "Trying It On for Size: How Does Design Fit into International Public Policy?" *Design Issues* 27, no. 2 (2011): 6–16.

Miller, Katherine. *Organizational Communication: Approaches and Processes*, 6th edition. Boston, MA: Wadsworth, 2006.

Mirivel, Julien C., and Karen Tracy. "Premeeting Talk: An Organizationally Crucial Form of Talk." *Research on Language and Social Interaction* 38 (2005): 1–34. doi:10.1207/s15327973rlsi3801_1.

Molina-Markham, Elizabeth. "Finding the 'Sense of the Meeting': Decision Making Through Silence among Quakers." *Western Journal of Communication* 78, no. 2 (2014): 155–174.

Molina-Markham, Elizabeth, Brion van Over, Sunny Lie, and Donal Carbaugh. "'You Can Do It Baby': Non-Task Talk with an In-Car Speech Enabled System." Manuscript submitted for publication, 2014.

Molinsky, Andrew L. "The Psychological Processes of Cultural Retooling." *Academy of Management Journal* 56, no. 3 (2013): 683–710. doi:10.5465/amj.2010.0492.

Moneta, Giovanni B., and Synnöve S. Kekkonen-Moneta. "Affective Learning in Online Multimedia and Lecture Versions of an Introductory Computing Course." *Educational Psychology* 27, no. 1 (2007): 51–74. doi:10.1080/01443410601061413.

Moon, Jennifer A. *A Handbook of Reflective and Experiential Learning*. New York: Routledge Falmer, 2004.

Murphy, Elizabeth. "Recognising and Promoting Collaboration in an Online Asynchronous Discussion." *British Journal of Educational Technology* 35, no. 4 (2004): 421–31. doi:10.1111/j.0007-1013.2004.00401.x.

Murphy, Elizabeth, María A. Rodríguez-Manzanares, and Michael Barbour. "Asynchronous and Synchronous Online Teaching: Perspectives of Canadian High School Distance Education Teachers." *British Journal of Educational Technology* 42, no. 4 (2011): 583–91. doi:10.1111/j.1467-8535.2010.01112.x.

Murray, Janet H. *Inventing the Medium: Principles of Interaction Design as a Cultural Practice*. Cambridge, MA: The MIT Press, 2012.

Nass, Clifford Ivar, and Scott Brave. *Wired for Speech: How Voice Activates and Advances the Human-Computer Relationship*. Cambridge, MA: MIT Press, 2005.

Nass, Clifford, Jonathan Steuer, and Ellen R. Tauber. "Computers Are Social Actors." Paper presented at the CHI, 1994.

Nass, Clifford Ivar, and Corina Yen. *The Man Who Lied to His Laptop: What Machines Teach Us about Human Relationships*. New York: Current, 2010.

NCDAE. "Tips and Tools: Content Management Systems and Accessibility." *Content Management Systems & Accessibility* (June 2006). Accessed January 22, 2015. http://ncdae.org/resources/factsheets/cms.php.

Nelson, Edwin B. "Cultural Awareness: Resources Can Help Prepare Soldiers before Deployments." *Infantry Magazine*, January/February 2007.

Nickols, Fred. "The Knowledge in Knowledge Management." In *The Knowledge Management Yearbook 2000–2001*, edited by John A. Woods and James Cortada, 12–21. Boston, MA: Butterworth-Heineman, 2000.

Nielsen, Jakob. "Heuristic Evaluation." In *Usability Inspection Methods*, edited by Jakob Nielsen and Robert L. Mack, 25–62. New York: John Wiley & Sons, Inc., 1994.

Nielsen, Jakob. "10 Usability Heuristics for User Interface Design." Accessed January 2, 2015. http://www.nngroup.com/articles/ten-usability-heuristics/.

Nieminen, Hannu. "Medioituminen ja suomalaisen viestintämaiseman muutos" [Mediazation and the Change in Finnish Communication Scene]. In *Uusi media ja arkielämä. Kirjoituksia uuden ajan kulttuurista*, Publications A, No. 41, edited by Hannu Nieminen, Petri Saarikoski, and Jaakko Suominen, 18–43. Turku: University of Turku, School of Art Studies, 2000.

Norman, Don. *The Design of Everyday Things*. New York: Basic Books, 2002.

Official Statistics of Finland (OSF). "Use of Information and Communications Technology by Individuals" [e-publication]. Helsinki: Statistics Finland, 2014. Accessed January 15, 2015. http://www.stat.fi/til/sutivi/2014/sutivi_2014_2014-11-06_tie_001_en.html.

Omar, Alwiya S. "Closing Kiswahili Conversations: The Performance of Native and Non-native Speakers." *Pragmatics and Language Learning* 4 (1993): 104–125.

Pacanowsky, Michael E., and Nick O'Donnell-Trujillo. "Organizational Communication as Cultural Performance." *Communication Monographs* 50, no. 2 (1983): 126–47. doi:10.1080/03637758309390158.

Pan, Yuling, Suzanne W. Scollon, and Ron Scollon. *Professional Communication in International Settings*. Malden, MA: Blackwell, 2002.

PCW. "Adopting an Agile Methodology Requirements—Gathering and Delivery" (2014). Accessed February 6, 2015. https://www.pwc.com/en_US/us/insurance/publications/assets/pwc-adopting-agile-methodology.pdf, retrieved 2/6/2015.

Peltonen, Matti. "Omakuvamme murroskohdat. Maisema ja kieli suomalaisuuskäsityksen perusaineksina" [Turning-Points of Our Self-Image. Landscape and Language as the Basic Ingredients of the Conception of Finnishness]. In *Elävänä Euroopassa. Muuttuva suomalainen identiteetti*, edited by Pertti Alasuutari and Petri Ruuska, 19–40. Tampere: Vastapaino, 1998.

Philipsen, Gerry. "Speaking 'Like a Man' in Teamsterville: Culture Patterns of Role Enactment in an Urban Neighborhood." *Quarterly Journal of Speech* 61, no. 1 (1975): 13–23.

Philipsen, Gerry. "The Prospect for Cultural Communication." In *Communication Theory: Eastern and Western Perspectives*, edited by Lawrence D. Kincaid, 245–54. Orlando, FL: Academic Press, 1987.

Philipsen, Gerry. *Speaking Culturally: Explorations in Social Communication*. Albany, NY: State University of New York Press, 1992.

Philipsen, Gerry. "A Theory of Speech Codes." In *Developing Communication Theories*, edited by Gerry Philipsen and Terrance L. Albrecht, 119–56. New York: State University of New York Press, 1997.

Philipsen, Gerry. "Permission to Speak the Discourse of Difference: A Case Study." *Research on Language & Social Interaction* 33, no. 2 (2000): 213–34.

Philipsen, Gerry. "Cultural Communication." In *Handbook of International and Intercultural Communication*, edited by William B. Gudykunst and Bella Mody, 51–67. Thousand Oaks, CA: Sage, 2002.

Philipsen, Gerry. "Some Thoughts on How to Approach Finding One's Feet in Unfamiliar Cultural Terrain." *Communication Monographs* 77, no. 2 (2010): 160–68.

Philipsen, Gerry. "Local Strategies Research: From Knowledge to Practice." Unpublished paper, 2012.

Philipsen, Gerry, and Lisa M. Coutu. "The Ethnography of Speaking." In *Handbook of Language and Social Interaction*, edited by Kristine L. Fitch and Robert E. Sanders, 355–79. Mahwah, NJ: Lawrence Erlbaum Associates, 2005.

Philipsen, Gerry, Lisa M. Coutu, and Patricia Covarrubias. "Speech Codes Theory: Restatement, Revisions, and Response to Criticisms." In *Theorizing About Intercultural Communication*, edited by William Gudykunst, 55–68. Thousand Oaks, CA: Sage, 2005.

Philipsen, Gerry, and James L. Leighter. "Sam Steinberg's Use of 'Tell' in After Mr. Sam." In *Interacting and Organizing: Analyses of a Management Meeting*, edited by Francois Cooren, 205–23: Lawrence Erlbaum Associates, Inc., 2007.

Picciano, Anthony G. "Beyond Student Perceptions: Issues of Interaction, Presence, and Performance in an Online Course." *Journal of Asynchronous Learning Networks* 6, no. 1 (2002): 21–40. doi:10.1.1.98.6506.

Potter, Jonathan, and Margaret Wetherell. *Discourse and Social Psychology: Beyond Attitudes and Behaviour.* Thousand Oaks, CA: Sage, 1987.

Poyner, Robert D. "The Hyphenated Airman." *Air and Space Power Journal* 21, no. 3 (Fall 2007): 23–25. Accessed February 6, 2015. http://www.au.af.mil/au/afri/aspj/airchronicles/apj/apj07/fal07/fal07.pdf.

Rempel, Hannah G., and Paula S. McMillan. "Using Courseware Discussion Boards to Engage Graduate Students in Online Library Workshops." *Internet Reference Services Quarterly* 13, no. 4 (2008): 363–80. doi:10.1080/10875300802326350.

Restauri, Sherri L. "Creating an Effective Online Distance Education Program Using Targeted Support Factors." *TechTrends* 48, no. 6 (2004): 32–39. doi:10.1007/BF02763580.

Rice, Ronald E., and Paul M. Leonardi. "Information and Communication Technologies in Organizations." In *Sage Handbook of Organizational Communication*, 3rd edition, edited by Linda Putnam and Dennis Mumby, 425–488. Thousand Oaks, CA: Sage, 2013.

Rivait, Jessica. "Course Management Systems as Ongoing Classroom Memory: ANGEL as a Living Archive." *Kairos: A Journal of Rhetoric* 15, no. 1 (July 11, 2014). Accessed February 6, 2015. http://praxis.technorhetoric.net/tiki-index.php?page=Classroom_Memory.

Roberts, Barbara. "Interaction, Reflection and Learning at a Distance." *Open Learning: The Journal of Open and Distance Learning* 17, no. 1 (2002): 39-55. doi:10.1080/02680510120110166.

Rockstuhl, Thomas, Soon Ang, Kok-Yee Ng, Filip Lievens, and Linn VanDyne. "Putting Judging Situations into Situational Judgment Tests: Evidence from Intercultural Multimedia SJTs." *Journal of Applied Psychology* (October 6, 2014). Accessed February 6, 2015. http://dx.doi.org/10.1037/a0038098.

Rockstuhl, Thomas, A. Presbitero, Kok-Yee Ng, and Soon Ang. "Metacognitive Cultural Intelligence and Offshoring Performance: Predictive Validity of an Intercultural Situational Judgment Test (iSJT)." Presentation at *The 3rd International Conference on Outsourcing of Information Services*. Proceedings, Mannheim, Germany, June 10–11, 2013.

Rogers, Yvonne, Helen Sharp, and Jenny Preece. *Interaction Design: Beyond Human-Computer Interaction*. West Sussex, UK: John Wiley & Sons, 2011.

Rowley, Jennifer, and Jennifer O'Dea. "How Do Students Perceive the Enhancement of Their Own Learning? A Comparison of Two Education Faculties' Experiences in Building an Online Learning Community for Bachelor of Music Education and Bachelor of Education Students." In *Australian Teacher Education Association Annual Conference Proceedings Archive*. Proceedings of "Teacher Education Crossing Borders: Cultures, Contexts, Communities and Curriculum" the Annual Conference of the Australian Teacher Education Association (ATEA), Albury. (July 2009). Accessed February 6, 2015. http://atea.edu.au/ConfPapers/2009/Refereed/Rowley.pdf.

Ruud, Gary. "The Symbolic Construction of Organizational Identities and Community in a Regional Symphony." *Communication Studies* 46, no. 3–4 (1995): 201–221. doi:10.1080/10510979509368452.

Ruud, Gary. "The Symphony: Organizational Discourse and the Symbolic Tensions between Artistic and Business Ideologies." *Journal of Applied Communication Research* 28, no. 2 (2000): 117–143. doi:10.1080/00909880009365559.

Ruuska, Petri. "Negatiivinen suomalaisuuspuhe yhteisön rakentajana" [Negative Finnishness Talk Building a Community]. *Sosiologia* 36 (1999): 293–305.

Ruuska, Petri. *Kuviteltu Suomi. Globalisaation, nationalismin ja suomalaisuuden punos julkisissa sanoissa 1980–90-luvuilla* [Imagined Finland: Interweaving Globalization, Nationalism and Finnishness in Public Discourse in the 1980s and 1990s]. Acta Electronica Universitatis Tamperensis No. 156. Tampere: Tampereen yliopistopaino, 2002.

Sacks, Harvey. *Lectures in Conversation*, volumes 1–2. Malden, MA: Blackwell, 1992.

Sacks, Harvey. Lecture 2(R) "The Baby Cried. The Mommy Picked It Up" (ctd). In *Lectures on Conversation*, edited by Gail Jefferson and Emanuel L. Schegloff, 223. John Wiley & Sons, 1995.

Sacks, Harvey, Emanuel Schegloff, and Gail Jefferson. "A Simplest Systematics for the Organization of Turn-Taking for Conversation." In *Studies in the Organization of Conversational Interaction*, edited by Jim Schenkein, 7–55. New York: Academic Press, 1978.

Sandars, John, and Michelle Langlois. "Online Learning Networks for General Practitioners: Evaluation of a Pilot Project." *Education for Primary Care* 16 (2005): 688–96.

Sandel, Todd L. "Narrated Relationships: Mothers-in-Law and Daughters-in-Law Justifying Conflicts in Taiwan's Chhan-chng." *Research on Language and Social Interaction* 37, no. 3 (2004): 365–398.

Sandel, Todd. "'Oh, I'm Here!': Social Media's Impact on the Cross-Cultural Adaptation of Students Studying Abroad." *Journal of Intercultural Communication Research* 43, no. 1 (2014): 1–29. doi:10.1080/17475759.2013.865662.

Sandel, Todd L. *Brides on Sale: Taiwanese Cross Border Marriages in a Globalizing Asia.* New York: Peter Lang, 2015.

Santovec, Mary Lou. "Doing Online Professional Development—Online." *Distance Education Report* (September 15, 2004): 6–7.

Saville-Troike, Muriel. *The Ethnography of Communication: An Introduction.* Baltimore, MD: University Park Press, 1982.

Saville-Troike, Muriel. *The Ethnography of Communication: An Introduction*, 3rd edition. Malden, MA: Blackwell Publishing, 2003.

Schegloff, Emanuel, and Harvey Sacks. "Opening Up Closings." *Semiotica* 8, no. 4 (1973): 289–327.

Schön, Donald A. *Educating the Reflective Practitioner: Towards a New Design For Teaching and Learning in the Professions.* San Francisco: Jossey-Bass, 1987.

Schön, Donald A. *The Reflective Practitioner.* San Francisco: Jossey-Bass, 1992.

Schwartzman, Helen B. *The Meeting: Gatherings in Organizations and Communities.* New York: Plenum Press, 1989.

Scollo, Michelle. "Cultural Approaches to Discourse Analysis: A Theoretical and Methodological Conversation with Special Focus on Donal Carbaugh's Cultural Discourse Theory." *Journal of Multicultural Discourses* 6, no. 1 (2011): 1–32.

Scollon, Ronald, and Suzanne Scollon. *Narrative, Literacy and Face in Interethnic Communication* 7. Norwood, NJ: Ablex Publishing Corporation, 1981.

Scott, Marvin B., and Stanford M. Lyman. "Accounts." *American Sociological Review* 33, no. 1 (1968): 46–62. Accessed February 6, 2015. http://links.jstor.org/sici?sici=0003-1224%28196802%2933%3A1%3C46%3AA%3E2.0.CO%3B2-M.

Searle, John. *Speech Acts: An Essay in the Philosophy of Language.* Cambridge, UK: Cambridge University Press, 1969.

Seargeant, Philip, and Caroline Tagg. "Introduction: The Language of Social Media." In *The Language of Social Media: Identity and Community on the Internet*, edited by Philip Seargeant and Caroline Tagg, 1–20. New York: Palgrave MacMillan, 2014.

Seargeant, Philip, Caroline Tagg, and Wipapan Ngampramuan. "Language Choice and Addressivity Strategies in Thai-English Social Network Interactions." *Journal of Sociolinguistics* 16, no. 4 (2012): 510–531.

Sellen, Abigail J. "Speech Patterns in Video-Mediated Conversations." In *CHI '92 Proceedings of the SIGCHI Conference on Human Factors in Computing Systems.* New York: ACM, 1992. doi:10.1145/142750.142756.

Sellen, Abigail, Yvonne Rogers, Richard Harper, and Tom Rodden. "Reflecting Human Values in the Digital Age." *Communications of the ACM* 52, no. 3 (2009): 58–66. doi:10.1145/1467247.1467265.

Shoemaker, Stowe. "Scripts: Precursor of Consumer Expectations." *Cornell Hotel and Restaurant Administration Quarterly* (February 1996): 42–53.

Short, Nancy M. "Asynchronous Distance Education." *Technological Horizons in Education Journal* 28, no. 2 (2000). Accessed February 6, 2015. http://thejournal.com/magazine/vault/A3001.cfm.

Shuter, Robert. "Introduction: New Media Across Cultures—Prospect and Promise." *Journal of International and Intercultural Communication* 4, no. 4 (2011): 241–245.

Sigman, Stuart. "Order and Continuity in Human Relationships: A Social Communication Approach to Defining 'Relationship.'" In *Social Approaches to Communication*, edited by Wendy Leeds-Hurwitz, 188–200. New York: Guilford, 1995.

Snow, Eleanour, and Perry Sampson. Course Platforms for Teaching Online. Authored as Part of the 2010 Workshop, Teaching Geoscience Online—A Workshop for Digital Faculty, 2010. Accessed February 6, 2015. http://serc.carleton.edu/NAGTWorkshops/online/platforms.html.

Spencer, Lyle M., and Signe M. Spencer. *Competence at Work: Models for Superior Performance.* New York: Wiley, 1993.

Sprain, Leah, and David Boromisza-Habashi. "Meetings: A Cultural Perspective." *Journal of Multicultural Discourses* 7, no. 2 (2012): 179–189. doi:10.1080/17447143.2012.685743.

Sprain, Leah, and David Boromisza-Habashi. "The Ethnographer of Communication at the Table: Building Cultural Competence, Designing Strategic Action." *Journal of Applied Communication Research* 41, no. 2 (2013): 181–187.

Sprain, Leah, and John Gastil. "What Does It Mean to Deliberate? An Interpretive Account of Jurors' Expressed Deliberative Rules and Premises." *Communication Quarterly* 61, no. 2 (2013): 151–71.

Staske, Shirley. "Claiming Individualized Knowledge of a Conversational Partner." *Research on Language & Social Interaction* 35, no. 3 (2002): 249–276. doi:10.1207/S15327973RLSI3503_1.

Sternberg, Janet. "Misbehavior in Mediated Places: Situational Proprieties and Communication Environments." *ETC: A Review of General Semantics* 66, no. 4 (2009): 433–22.

Suchman, Lucy. *Plans and Situated Actions: The Problem of Human-Machine Communication.* New York: Cambridge University Press, 1987.

Swan, Karen. "Learning Effectiveness: What the Research Tells Us." In *Elements of Quality Online Education: Practice and Direction*, edited by John R. Bourne and Janet C. Moore, 13–47. Needham, MA: Sloan Consortium, 2003.

Szabody, Gina, Crystal N. Fodrey, and Celeste Del Russo. "Digital (Re)Visions: Blening Pedagogical Strategies with Dynamic Classroom Tactics." *Kairos* 19.1 (Fall 2014).

Takami, Tomoko. "A Study on Closing Sections of Japanese Telephone Conversations." *Working Papers in Educational Linguistics* 18, no. 1 (2002): 67–85.

Talbot, David. "China's Internet Paradox." *MIT Technology Review* (April 14, 2010). Accessed February 6, 2015. http://www.technologyreview.com/featured-story/418448/chinas-internet-paradox/.

Teasley, Martell L. "Perceived Levels Of Cultural Competence Through Social Work Education And Professional Development For Urban School Social Workers." *Journal of Social Work Education* 41, no. 1 (2005): 85–98. doi:10.5175/JSWE.2005.200300351.

Toiskallio, Kalle, Sakari Tamminen, Heini Korpilahti, Salla Hari, and Marko Nieminen. *Mobiilit käyttökontekstit—Mobix. Loppuraportti* [Mobile Contexts of

Use—Mobix. Final Report]. Software Business and Engineering Institute. Helsinki University of Technology, 2004.

Topping, Keith. "Peer Assessment between Students in Colleges and Universities." *Review of Educational Research* 68, no. 3 (1998): 249–276.

Townsend, Rebecca. "Town Meeting as a Communication Event: Democracy's Act Sequence." *Research on Language and Social Interaction* 42, no. 1 (2009): 68–89.

Tracy, Karen, and Aaron Dimock. "Meetings: Discursive Sites for Building and Fragmenting Community." In *Communication Yearbook*, volume 28, edited by Pamela J. Kalbfleisch, 127–166. Mahwah, NJ: Lawrence Erlbaum Associates, 2004.

Tracy, Sarah J., Karen K. Myers, and Clifton W. Scott. "Cracking Jokes and Crafting Selves: Sensemaking and Identity Management among Human Service Workers." *Communication Monographs* 73, no. 3 (2006): 283–308. doi:10.1080/03637750600889500.

Tsimhoni, Omer, Ute Winter, and Timothy Grost. "Cultural Considerations for the Design of Automotive Speech Applications." Paper presented at the 17th World Congress on Ergonomics IEA, Beijing, China, 2009.

Turkle, Sherry. *Alone Together: Why We Expect More from Technology and Less from Each Other.* New York: Basic Books, 2011.

Van De Ven, Andrew H. *Engaged Scholarship: A Guide for Organizational & Social Research.* Oxford: Oxford University Press, 2007.

van Over, Brion. "Tracing the Decay of a Communication Event: The Case of *The Daily Show's* 'Seat of Heat.'" *Text & Talk* 34, no. 2 (2014): 187–208.

Varpio, Yrjö. *Pohjantähden maa. Johdatusta Suomen kirjallisuuteen ja kulttuuriin* [Country of the North Star. Introduction to Finnish Literature and Culture]. Tampere: Tampere University Press, 1999.

Varvel, Todd K. "Ensuring Diversity Is Not Just Another Buzz Word." Master's thesis, Air University, 2000. Accessed February 6, 2015. http://www.au.af.mil/au/awc/awcgate/acsc/00-180.pdf.

Vehviläinen, Marja. "Teknologinen nationalismi" [Technological Nationalism]. In *Suomineitonen hei! Kansallisuuden sukupuoli*, edited by Tuula Gordon, Katri Komulainen, and Kirsti Lempiäinen, 211–29. Tampere: Vastapaino, 2002.

Vorvoreanu, Mihaela. "Perceptions of Corporations on Facebook: An Analysis of Facebook Social Norms." *Journal of New Communications Research* 4, no. 1 (2009): 67–86.

Walther, Joseph B. "Interpersonal Effects in Computer-Mediated Interaction: A Relational Perspective." *Communication Research* 19, no. 1 (1992): 52–90.

Walther, Joseph B. "Interaction through Technological Lenses: Computer-Mediated Communication and Language." *Journal of Language and Social Psychology* 31, no. 4 (2012): 397–414.

Weingardt, Kenneth R., Michael A. Cucciare, Christine Bellotti, and Wen Pin Lai. "A Randomized Trial Comparing Two Models of Web-based Training in Cognitive–Behavioral Therapy for Substance Abuse Counselors." *Journal of Substance Abuse Treatment* 37, no. 3 (2009): 219–27. doi:10.1016/j.jsat.2009.01.002.

Wenger, Etienne. "Brief Introduction to Communities of Practice" (n.d.). Accessed February 9, 2015. http://wenger-trayner.com/wp-content/uploads/2013/10/06-Brief-introduction-to-communities-of-practice.pdf.

Whittaker, Steve, and Brid O'Conaill. "The Role of Vision in Face-to-Face and Mediated Communication." In *Video-Mediated Communication*, edited by Kathleen E. Finn, Abigail J. Sellen, and Sylvia B. Wilbur, 23–49. Mahwah, NJ: Lawrence Erlbaum Associates, 1997.

Wieder, Lawrence, and Steven Pratt. "On Being a Recognizable Indian among Indians." In *Cultural Communication and Intercultural Contact*, edited by Donal Carbaugh, 45–64. Hillsdale, NJ: Lawrence Erlbaum, 1990.

Wiio, Osmo A. "Matkapuhelimen voittokulku" [Triumphal March of the Mobile Phone]. Review of *Alkuräjähdys — Radiolinja Suomen GSM-matkapuhelintoiminta 1988–1998*, by Martti Häikiö. *Kanava* 8 (1998): 513–5.

Wilkins, Richard. "'Asia' (Matter-of-Fact) Communication: Finnish Cultural Terms for Talk in Education Scenes." PhD dissertation, University of Massachusetts at Amherst, 1999.

Wilson, Suzanne, Robert E. Floden, and Joan Ferrini-Mundy. *Teacher Preparation Research: Current Knowledge, Gaps, and Recommendations.* Seattle, WA: Center for the Study of Teaching and Policy, 2001.

Winchatz, Michaela R. "Social Meanings in German Interactions: An Ethnographic Analysis of the Second-Person Pronoun Sie." *Research on Language and Social Interaction* 34, no. 3 (2001): 337–69.

Winter, Ute, Yael Shmueli, and Timothy Grost. "Interaction Styles in Use of Automotive Interfaces." In *Proceedings of the Afeka AVIOS 2013 Speech Processing Conference,* Tel Aviv, Israel, 2013.

Winter, Ute, Omer Tsimhoni, and Timothy Grost. "Identifying Cultural Aspects in Use of In-Vehicle Speech Applications." In *Proceedings of the Afeka AVIOS 2011 Speech Processing Conference,* Tel-Aviv, Israel: Literature, 2011.

Witteborn, Saskia. "Communicative Competence Revisited: An Emic Approach to Studying Intercultural Communicative Competence." *Journal of Intercultural Communication Research* 32, no. 3 (2003): 187–203.

Witteborn, Saskia. "Discursive Grouping in a Virtual Forum: Dialogue, Difference, and the 'Intercultural.'" *Journal of International and Intercultural Communication* 4, no. 2 (2011): 109–26. doi:10.1080/17513057.2011.556827.

Witteborn, Saskia. "Forced Migrants, New Media Practices, and the Creation of Locality." In *The Handbook of Global Media Research*, edited by Ingrid Volkmer, 312–30. Malden, MA: Blackwell Publishing Ltd., 2012.

Witteborn, Saskia, Trudy Milburn, and Evelyn Y. Ho. "The Ethnography of Communication as Applied Methodology: Insights from Three Case Studies." *Journal of Applied Communication Research* 41, no. 2 (2013): 188–94.

Witteborn, Saskia, and Leah Sprain. "Grouping Processes in a Public Meeting from an Ethnography of Communication and Cultural Discourse Analysis Perspective." *The International Journal of Public Participation* 3, no. 2 (2009): 14–35.

Wodak, Ruth, Rudolf de Cillia, Martin Reisigl and Karin Liebhart. *The Discursive Construction of National Identity.* Edinburgh: Edinburgh University Press, 2009.

Wolcott, Harry F. *Ethnography: A Way of Seeing.* Walnut Creek, CA: AltaMira Press, 1999.

Woodward, Helen. "Reflective Journals and Portfolios: Learning through Assessment." *Assessment & Evaluation in Higher Education* 23 (1998): 415–423.

Wu, Si. *I am Not Fighting Alone: The Impact of Social Media on the Cross-Border Study Experience of Mainland Students in Macau.* Macao, SAR: Master's thesis, University of Macau, 2014.

Xia, Jianhong, John Fielder, and Lou Siragusa. "Achieving Better Peer Interaction in Online Discussion Forums: A Reflective Practitioner Case Study." *Issues in Educational Research* 23, no. 1 (2013): 97–113. February 6, 2015. http://www.iier.org.au/iier23/xia.pdf.

Yamada, Haru. "Topic Management and Turn Distribution in Business Meetings: American versus Japanese Strategies." *Text—Interdisciplinary Journal for the Study of Discourse* 10, no. 3 (1990). doi:10.1515/text.1.1990.10.3.271.

Yamada, Haru. *American and Japanese Business Discourse: A Comparison of Interactional Styles.* Norwood, NJ: Ablex, 1992.

Yamada, Haru. "Organisation in American and Japanese Meetings: Task versus Relationship." In *The Language of Business: An International Perspective*, edited by Francesca Bargiela-Chiappini and Sandra J. Harris, 117–135. Edinburgh, UK: Edinburgh University Press, 1997.

Yang, Guobin. *The Power of the Internet in China.* New York: Columbia University Press, 2009.

Zeichner, Ken. "Rethinking the Connections between Campus Courses and Field Experiences in College- and University-Based Teacher Education." *Journal of Teacher Education* 61 (2010): 89–99.

Zha, Shenghua, and Christy L. Ottendorfer. "Effects of Peer-Led Online Asynchronous Discussion on Undergraduate Students' Cognitive Achievement." *American Journal of Distance Education* 25, no. 4 (2011). doi: DOI:10.1080/08923647.2011.618314.

Zubizarretta, John. *The Learning Portfolio: Reflective Practice for Improving Student Learning.* San Francisco: Jossey-Bass, 2009.

Index

About the Editor and Contributors

EDITOR

Trudy Milburn, PhD, is director of campus solutions at Taskstream, a cloud-based assessment software company located in New York City. She has been an associate scholar with the Center for Local Strategies Research, University of Washington, and participates in Teacher's College LANSI working group. Dr. Milburn is the co-author of *Citizen Discourse on Contaminated Water, Superfund Cleanups, and Landscape Restoration: (Re)making Milltown, Montana* and the author of *Nonprofit Organizations: Creating Membership through Communication*. Her work has appeared in publications such as *Journal of Applied Communication Research*, *Communication Monographs*, and *Business Communication Quarterly*. She has been a tenured associate professor on the faculties of California State University, Channel Islands, and Baruch College/The City University of New York, and is a past chair of the Language and Social Interaction Division of the National Communication Association.

CONTRIBUTORS

Maaike Bouwmeester, PhD, is former vice president of knowledge management at Taskstream and clinical assistant professor of educational communications and technology at New York University. She has spent the better part of two decades working at the intersection of educational software and learning theory. Her work focuses on educational platforms that support competency and outcomes based learning environments, and tools to support reflection, collaboration, e-portfolios, performance assessment, and educational quality.

Donal Carbaugh, PhD, is professor of communication at the University of Massachusetts–Amherst. His recent book is *Reporting Cultures on 60 Minutes: "That's not Funny"* (Routledge, 2015). His interests are in developing a culturally sensitive perspective on communication, persons, and society with special applications to dialogue, spirit, and nature.

Tabitha Hart, PhD, is assistant professor in the Department of Communication Studies at San Jose State University. Dr. Hart is the author of "Technologies for Conducting an Online Ethnography of Communication: The Case of Eloqi" (book chapter) and "The Interface is the Message: How a Technological Platform Shapes Communication in an Online Chinese & American Community" (journal article). Her research foci are cultural communication and applied strategic communication in online and offline organizational environments. You can learn more about her work by visiting http://tabithahart.net/.

Jenny Bei Ju, MA, is lecturer of the international business faculty at Beijing Normal University, Zhuhai Campus. She is also a doctoral student at the University of Macau. Ms. Ju's work has appeared in the *International Journal of Business and Management* and the *Journal of Jiangxi Administration Institute*. She is interested in studying social media and acculturation in China.

Sunny Lie, PhD, is assistant professor at Saint Cloud State University in Saint Cloud, MN. Her work has appeared in publications such as the *Journal of Applied Communication Research*, *Journal of Linguistic Anthropology*, *China Media Research*, and *Intercultural Pragmatics*. Her current research interest is the intersection between religion and identity as enacted through cultural communication, specifically U.S. American Buddhism and Chinese Indonesian Christianity.

James (Jay) L. Leighter, PhD, is associate professor in the Department of Communication Studies at Creighton University. He investigates how moments for speaking are culturally influenced and, in instances of planned intervention (for example, design, public health, social justice), what needs to be known, culturally and communicatively, about the community who will be affected by the intervention. His work has appeared in the *Journal of Applied Communication Research*, *Western Journal of Communication*, and the *International Journal of Participation*.

Lauren Mackenzie, PhD, joined the Bloomsburg University communication studies faculty in the fall of 2014 and teaches courses devoted to intercultural communication and conflict management. From 2009 to 2014, Dr. Mackenzie served as associate professor of cross-cultural communication at the U. S. Air Force Culture and Language Center. Recently, she has published in the *Journal of Culture, Language & International Security* and EDU-CAUSE *Review*, as well as the edited volume *Cross-Cultural Competence for a 21st Century Military: Culture, the Flipside of COIN*. Her research focuses on the verbal, paralinguistic, and nonverbal components that comprise the communication of respect across cultures.

Elizabeth Molina-Markham, PhD (University of Massachusetts–Amherst), has been published in periodicals such as the *Western Journal of Communication*, the *Journal of Communication and Religion*, the *Journal of Applied Communication Research*, and *Narrative Inquiry*. Her research focuses on cultural communication practices, in particular among members of the Religious Society of Friends (Quakers) and also between drivers and speech-enabled technology.

Katherine R. Peters, MA, is a doctoral student in the Department of Communication at the University of Colorado Boulder. Her research is focused on the cultural foundations of organizing, with a particular focus on meetings. She is also interested in how technologies and communication practices, such as meetings, enable and constrain organizing.

Saila Poutiainen, PhD, is a tenured university lecturer of speech communication in the Institute of Behavioural Sciences, University of Helsinki, Finland. Dr. Poutiainen has recently edited *Theoretical Turbulence in Intercultural Communication Studies*. She serves as a board member in the Nordic Network for Intercultural Communication and for the Intercultural Encounters Master's program at the University of Helsinki.

Todd Sandel, PhD, is associate professor of communication at the University of Macau, Macao SAR, China. He is author of *Foreign Brides on Sale: Taiwanese Women's and Men's Cross Border Marriages in a Globalizing Asia*, and associate editor of *The International Encyclopedia of Language and Social Interaction*. Dr. Sandel's work has appeared in publications such as *China Media Research, Journal of Contemporary China, Journal of Intercultural Communication Research, Journal of International and Intercultural Communication, Language in Society, Research on Language and Social Interaction*, and *Social Development*. He is a past chair of the Language and Social Interaction Division of NCA.

Megan R. Wallace is an analyst with Booz Allen Hamilton at the Air Force Culture and Language Center. Ms. Wallace has co-authored publications in *Academic Exchange Quarterly* and *Cross-Cultural Communication* and was a co-contributor to the recent edited volume *Cross-Cultural Competence for a 21st Century Military: Culture, the Flipside of COIN.* During her career, she has provided instructional support to the Community College of the Air Force, the University of South Alabama, and Park University.

Brion van Over, PhD, is an instructor of communication at Manchester Community College, Manchester, Connecticut, and graduate of the University of Massachusetts–Amherst. Dr. van Over's work has appeared in publications such as *Text & Talk*, *Journal of Applied Communication Research*, and *Journal of Pragmatics.* His research investigates the cultural constitution of communicative practices as shaped in discourse about place, identity, and communication itself.